Integral Dynamical Models
Singularities, Signals and Control

WORLD SCIENTIFIC SERIES ON NONLINEAR SCIENCE

Editor: Leon O. Chua
   University of California, Berkeley

---

*To view the complete list of the published volumes in the series, please visit:
http://www.worldscientific.com/series/wssnsa

WORLD SCIENTIFIC SERIES ON
NONLINEAR SCIENCE   Series A   Vol. 87
Series Editor: Leon O. Chua

Integral Dynamical Models
Singularities, Signals and Control

Denis Sidorov
Russian Academy of Sciences, Russia

World Scientific

NEW JERSEY · LONDON · SINGAPORE · BEIJING · SHANGHAI · HONG KONG · TAIPEI · CHENNAI

Published by

World Scientific Publishing Co. Pte. Ltd.
5 Toh Tuck Link, Singapore 596224
USA office: 27 Warren Street, Suite 401-402, Hackensack, NJ 07601
UK office: 57 Shelton Street, Covent Garden, London WC2H 9HE

Library of Congress Cataloging-in-Publication Data
Sidorov, Denis, 1974–   author.
   Integral dynamical models : singularities, signals and control / Denis Sidorov (Russian Academy of Sciences, Russia).
      pages cm -- (World scientific series on nonlinear science. Series A ; volume 87)
   Includes bibliographical references and index.
   ISBN 978-9814619189 (hardcover : alk. paper)
   1. Nonlinear integral equations. 2. Volterra equations. I. Title. II. Series: World Scientific series on nonlinear science. Series A, Monographs and treatises ; v. 87.
   QA431.S536 2014
   515'.45--dc23

                                                                2014029159

British Library Cataloguing-in-Publication Data
A catalogue record for this book is available from the British Library.

In-house Editor: Song Yu

In memory of Professor Alfredo Lorenzi

and Professor Vladilen Trenogin

Foreword

In order to progress in applying integral dynamical models to engineering and economic fields, one needs the understanding of applied problems as well as profound knowledge of continuous-time linear and nonlinear integral models. The aim of this text is to report recent advances in analytical and numerical techniques for the analysis, identification and control of a series of linear and nonlinear integral dynamical models. The applied mathematicians and the R & D engineers will find in this book new conceptions, tools and models for the solution of practical problems. The pure mathematicians and the theoretical physicists can see how the theory of functional equations, distributions, convex majorants, or fixed-point theorems can be employed to study linear and nonlinear equations in integral dynamical models. The main models addressed in this book involve novel classes of the Volterra integral equations of the first kind with jump discontinuous kernels and finite Volterra series models of dynamical systems. The principal message of this book is the systematic employment of integral operators in different mathematical models. The book illustrates the importance of integral operators and integral transforms in pure mathematics and, on this basis, in various direct and inverse problems in applied mathematics. This book is intended to help both postgraduates and practitioners understand the theory and application of integral dynamical models in electric power and heat engineering, economics and multidimensional signal processing.

Professor Hans-Jürgen Reinhardt

Universität Siegen
December, 2012

Contents

Part 2 Nonlinear Models, Singularities and Control 65

Chapter 1

Introduction and Overview

1.1 Statement of Purpose

The theory of integral equations, transforms and models is well developed
and dense in the literature and applications, since they have been found to
provide efficient ways of solving a variety of problems arising in engineering,
physical sciences and economics. However we have been unable to find a
systematic treatment of integral models of casual and evolving nonlinear
dynamical systems. Therefore this text is a modest attempt to partly fill
that gap.

The topic *dynamical systems* is the study of the long standing behavior of evolving systems, of which systematic theoretical studies have been
initiated at the end of XIX century concerning the fundamental questions
of the evolution and stability of the solar system. These studies led to the
development of diversified and powerful fields with various applications in
economics, ecology, biology, energetics, meteorology, astronomy, and other
areas.

By way of introduction let us consider the following integral equation

$$\int_a^b K(t,s)x(s)\,ds = f(t). \tag{1}$$

With known function $f(t)$ and kernel $K(t,s)$ this equation is known as the
Fredholm equation of the first kind with respect to $x(s)$. On the other hand,
the relation (1) can be considered as a general integral transform, e.g., we
can consider the following case $a = -\infty$, $b = +\infty$, $K(t,s) = \frac{1}{\sqrt{2\pi}}e^{-its}$, the
functions x and f are correspondingly input and output signals, we get as
is well known in digital signal processing, the Fourier integral transform

$$\int_{-\infty}^{+\infty} \frac{1}{\sqrt{2\pi}}e^{-its}x(s)\,ds = \hat{f}. \tag{2}$$

The solution of the equation (2) would give the inverse Fourier transform. Similarly we can define the Fourier cosine transform

$$\sqrt{\frac{2}{\pi}} \int_0^\infty \cos(ts)x(s)\,ds = f(t),$$

which is the well-known image and signal processing tool for lossy compression. The Laplace transform, Hilbert transform and other widely used integral transforms can be defined similarly.

If we consider the case of $a = -\infty$, $b = t$ we obtain a general representation for finite-dimension deterministic linear dynamical system which is well known in system control theory as the input-output explicit integral dynamical model with infinite delay

$$\int_{-\infty}^t K(t,s)x(s)\,ds = f(t),$$

where kernel matrix $K(t,s)$ is the impulse response function, $x(t)$ is input vector, $f(t)$ is vector of outputs. It is to be noted that many real-life input-output systems involve feedback. Most production and economic systems are also systems with memory. A real-life input-output dynamic system involves a feedback $A(t,s)$ that connects inputs x and outputs f, where A specifies the feedback type and intensity, then

$$\int_{-\infty}^t K(t,s)A(t,s,f(s))\,ds = f(t).$$

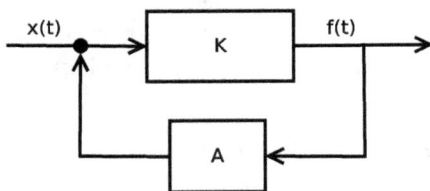

Fig. 1.1.   Feedback control.

These models take into account the memory/aftereffect of a dynamical system when its past impacts its future evolution. The memory is implemented in the existing technological and financial structure of physical capital (equipment). The memory duration is determined by the age of the oldest capital unit (e.g., equipment) still employed.

When applying these ideas to a problem in economics it is convenient to write $A(t,s)$ as $\Lambda(t,s)Y(s)$, where Y is distribution of products and the intensity control function $\Lambda(t, s)$ reflects the varying load of different equipment types in the production of various products. The intensity control Λ could be specific, e.g., it has to be capable to model the case when production systems use the newest, most efficient facilities and scrap the oldest when it becomes obsolete.

Taking into account the control function A, the input-output explicit integral dynamical model becomes the linear Volterra model

$$\int_{t_0}^{t} K(t, s)x(s)\, ds = f(t), \quad t \in [t_0, T] \tag{3}$$

with special kernels that have jump discontinuities on delay control curves. Here readers may refer to the article of [Hritonenko and Yatsenko (2005)] As we see now the Volterra (evolutionary) equations with jump discontinuous kernels play an important role in the theory of evolving dynamical systems in economics, ecology and energetics. The origins on the integral models studied in this book are from the class of the macroeconomic models with technical progress, integral models of [Solow (1969)] are examples of this kinds of models. Such kinds of models were initially proposed by [Kantorovich and Gor'kov (1959)]. [Kantorovich and Gor'kov (1959)] proposed a new integral macroeconomic model which is a development of the Solow model. Independently of Kantorovich, [Glushkov *et al.* (1983)] proposed a two-sector integral model described by a system of the Volterra equations which are connected with the Volterra equation of the first kind studied in this book. Volterra integral equations are used to model causal dynamical systems with memory. "Causal" essentially means that the output signal at each time instant t depends only on the history of the input signal up to time t. "Memory" means that the output at time t does not depend exclusively on the instant value of the input at time t, but rather it depends on the entire history of the input from some initial time t_0, where $t < t_0$.

In industrial applications and applications in other fields there is a large demand to apply nonlinear algorithms to control nonlinear dynamical systems with memory. For example, the evolution of an epidemic depends on the entire history of the epidemic; the number of individuals who will contract a disease at time t does not depend only on the number of infected individuals at time t, but also depend on the history of the infected numbers, because the disease conventionally progresses in several stages, and patients at different stages of infection have different symptoms. In economics the effectiveness of a strategy depends not only on the current state of the relevant economic system but also on the history of the system, e.g., the effectiveness of advertising, the adaptation of consumers to new products or new prices, and other relevant processes are not memoryless effects. Other examples of causal systems with memory include hereditary continuum mechanics, namely materials with memory, flow of water in porous media and other generally nonlinear systems. With algorithms considering the process nonlinearities, better control performance is expected in the whole operating range compared with conventional linear control algorithms. To cope with this demand the significant part of this book is devoted to the generalizations of the continuous-time Volterra models for casual nonlinear dynamical systems feed-back control. Such Volterra models appear from the generalization of the Volterra integral-functional series which is the conventional tool in nonlinear casual dynamical systems modelling introduced in classic book of [Volterra (2005)]. In the proposed generalized Volterra models the transfer functions $K_j(t, s_1, \ldots s_j, u(s_1), \ldots u(s_j))$ are assumed to be known. It is to be noted here that the Volterra series belongs to one of the best understood nonlinear system representations in signal processing. Namely we address the following nonlinear equation w.r.t. the continuous solution $u(t) \to u_0$ as $t \to 0$

$$\Phi\left(\int_0^t K_1(t, s, u(s))ds, \int_0^t \int_0^t K_2(t, s, s_1, s_2, u(s_1), u(s_2))ds_1 ds_2,\right.$$

$$\left.\ldots \int_0^t \cdots \int_0^t K_n(t, s_1, \ldots s_n, u(s_1), \ldots u(s_n))ds_1 \ldots ds_n, u(t), t\right) = 0,$$

where $t \in [0, T)$, and $\Phi : E_1 \times \cdots \times E_1 \times \mathbb{R}^1 \to E_2$ is nonlinear continuous operator of $n + 1$ variables $\omega_1, \ldots, \omega_n, u$, which are abstract continuous functions of real variable t with values in the Banach space E_1, $\Phi(0, \ldots, 0, u_0, 0) = 0$, $K_i : \underbrace{\mathbb{R}^1 \times \ldots \mathbb{R}^1}_{i+1} \times \underbrace{E_1 \times \cdots \times E_1}_{i} \to E_2$ are nonlinear continuous operators depending on $u(s) = (u(s_1), \ldots, u(s_n))$ and $t, s_1, \ldots s_n$ are real variables. Special attention is given to the "main" solution of such equations in sense of L. V. Kantorovich. Here readers may refer to classic book of [Kantorovich *et al.* (1950)].

A special case of such a nonlinear integral equation is the following polynomial equation with convolution kernels

$$\sum_{m=1}^N \int_0^t \cdots \int_0^t K_m(t - s_1, \ldots, t - s_m) \prod_{i=1}^m x(s_i)ds_i - f(t) = 0, \ 0 \le t \le T.$$

The method for construction of the generalized solutions (i.e., impulse control) to such an equation is also presented in this text. It is to be noted that such models are known in the signal processing community as continuous-time analogue to generalized polynomial kernel regression models, here readers may refer to book of [Mathews and Sicuranza (2000)] and to [Franz and Schölkopf (2006)]. The application of autoregressive models in power engineering for unstable oscillations detection is also studied in this text.

Among other nonlinear equations studied in this book is the following Hammerstein integral equation

$$u(x) = \int_a^b K(x,s)(u(s) + f(s,u,\lambda))\,ds,$$

and an abstract nonlinear Volterra equation with non-invertible operator. In this book we further develop and employ the theory of nonlinear parameter-dependent mathematical models introduced in book of [Sidorov *et al.* (2002)].

The proofs of the theorems on asymptotic behavior of solutions contain (rather simple) elements of functional analysis and were written in a way that one may omit the proofs since they are not of prejudice to the understanding of the main ideas of the book.

The principal part of the text is made up of various results that are mostly available in research papers, some in English or only in Russian. A number of results presented in this text have not appeared in print before, and previously published results have been substantially modified with examples and also improved.

1.2 The Scope of This Book

It is assumed that the reader has some background in integral and differential equations, functional analysis, some knowledge of digital signal processing is also expected. For some parts of the text the reader will need knowledge of the theory of distributions.

The book consists of three parts. Part 1 (Chapters 2–5) of this monograph presents the detailed overview of the theory behind the continuous-time Volterra models of evolving dynamical systems. For didactical reasons we start this book with single Volterra Integral Equations (VIEs), then we address systems of VIEs with piecewise continuous kernels. Finally we generalize our results on abstract Volterra operator equations with piecewise continuous kernels and consider solutions in the distributions space. Thereby, in Chapter 2 we introduce the VIEs of the first kind with piecewise continuous kernels, describe the structure of solutions and prove the existence theorem. In Chapters 3 and 4 we address a system of VIE with piecewise continuous kernels and the abstract integro-functional Volterra operator equation. In Chapter 5 we examine the special case when the VIE has no continuous solution and the generalized solutions are constructed in the Sobolev-Schwartz distributions space.

In Part 2 (Chapters 6–11) of the text we systematically study nonlinear continuous-time integral models starting from the Hammerstein integral equation, which is a well-known tool in control theory. In Chapter 7 we employ the similar technique (presented in Chapter 6) for the nonlinear Volterra equation of the second kind with non-invertible operator in the main part. In Chapter 8 solutions to nonlinear differential equations in the neighborhood of branching points are studied. In Chapters 9, 10 and 11 we correspondingly study the classical and generalized solutions to nonlinear Volterra equations arising in the theory of nonlinear continuous-time casual dynamical systems control based on the Volterra integral-functional series. Chapter 10 presents the theoretical results on generalized solution of the special class of nonlinear Volterra equations important for solution of the integral equations discussed in Chapters 9 and 11. In Section 11.3.1 we review state-of-the-art identification algorithms for Volterra models. The material of the book is organized so that as continuous-time methods for the Volterra models are presented, some corresponding discrete-time methods are also presented.

Part 3 (Chapters 12–14) is devoted to applications of the integral dynamical models. In Chapter 12 Volterra series based models are applied to nonlinear heat-exchange dynamics modelling. As applied to Subsection 12.2 we outline the elements of the economics theory of evolving dynamical systems, and describe the numerical method for solution of discrete-time VIEs of the first kind with piecewise continuous kernels. Although most of the examples and applications in this book are concerned with energetics and electrical systems, some examples and applications utilize dynamical systems from other fields, e.g., motion picture restoration. This is done to illustrate the diversity of applications for methods of signal processing and integral models. Indeed, Chapter 13 presents the practical example of using the Fourier transform in motion picture restoration. The book is concluded with Chapter 14 where the problem of electrical power systems parameters forecasting is addressed using the Hilbert integral transform and artificial neural networks. In Chapter 14, Subsection 14.2.3 we also discuss the unifying view on the Volterra theory and polynomial modelling as a kernel regression and present the recent results concerning application of non-stationary autoregressive models for on-line detection of inter-area oscillations in power systems.

1.3 Acknowledgments

This book benefitted from the support of the Russian Scientific Foundation, project No. 14-19-00054, the Russian Ministry of Education & Science Project "Singular Integral Models and Transforms: Theory & Applications" (State Contract No. 14.B37.21.0365), Seventh Framework EU Programme (ICOEUR Project), Sixth Framework EU Programme (BRAVA Project) under IST research and technological development programme and German Academic Exchange Service (DAAD) Grant No. A1200665, Russian Foundation of Basic Research Grant No. 11-08-00109. The author is grateful for discussions to Prof. Alfredo Lorenzi on the Volterra equations during his visit to the University of Milano and during the conference on "Inverse and Ill-Posed Problems of Mathematical Physics" in Novosibirsk. Also the author expresses his deepest gratitude to Prof. Hans-Jürgen Reinhardt for his hospitality during my stay in Siegen and to Prof.-Ing. Christian Rehtanz for kind invitation to present part of these lectures at the University of Dortmund during the research stay in Germany. The author is particularly grateful to Prof. Nikolai A. Sidorov and Prof. Vladilev A. Trenogin for their deep co-operation and friendly advice of long-standing. The author is indebted to his collaborators and colleagues Prof. Anatoly S. Apartsyn, Prof. Karen Egiazarian, Prof. Mihail V. Falaleev, Dr. Alexander V. Kiselev, Prof. Anil C. Kokaram, Prof. Victor G. Kurbatsky, Dr. Paul Leahy, Dr. Eugenia V. Markova, Mr. Daniil A. Panasetsky, Prof. Alexei V. Savvateev, Dr. Inna V. Sidler, Dr. Vaclav Šmídl, Prof. Valery S. Sizikov, Dr. Svetlana V. Solodusha, Mr. Vadim A. Spiryaev and Dr. Nikita V. Tomin who deserve much credit for these notes. The author gratefully acknowledges Prof. Malcolm Brown, Prof. Vladimir K. Gorbunov, Prof. Alfredo Lorenzi and Prof. Hans-Jürgen Reinhardt for valuable discussions of the manuscript.

I'm grateful to my wife, Aliona, and my children Lev Ryan and Alisa for their patience and love.

PART 1

Volterra Models of Evolving Dynamical Systems

Part Summary

Part 1 of this monograph presents the detailed overview of the theory behind the Volterra models of the evolving dynamical systems. In Chapter 2 we introduce the following Volterra Integral Equations (VIEs) of the first kind with piecewise continuous kernels

$$\int_0^t K(t,s)x(s)ds = f(t), \ \ 0 < t \le T,$$

$$K(t,s) = \begin{cases} K_1(t,s), & 0 \le s < \alpha_1(t), \\ K_2(t,s), & \alpha_1(t) \le s < \alpha_2(t), \\ \ \ldots & \ \ldots\ldots\ldots \\ K_n(t,s), & \alpha_{n-1}(t) \le s \le t, \end{cases}$$

with

$$0 < \alpha_1(t) < \alpha_2(t) < \cdots < \alpha_{n-1}(t) < t, \ |\alpha_i'(t)| < 1.$$

$K_i(t,s), i = \overline{1,n}$ and $f(t)$ are the continuous and sufficiently smooth functions, $f(0) = 0$. We describe the structure of the solution, a proof of existence of the solution is obtained. The special characteristic equation (see equation (2.4.1) for the scalar case) plays the principal role in the theory of such integral models. We analytically study the case when the characteristic equation has no integer or zero roots and the solution of the VIEs is unique. The numerical solution to the VIEs of the first kind with piecewise continuous kernels is discussed in Part 3. The method for the solution' asymptotic approximation construction is also suggested and the Main Theorem 3.1 is proved. Chapters 3 and 4 are devoted to generalizations of the results presented in Chapter 2 on the case of matrix VIE and in the case of abstract integro-functional Volterra operator equation. In the final Chapter 5 the special case when $f(0) \neq 0$ is addressed and generalized solution to the VIEs of the first kind with piecewise continuous kernel is constructed in the space of Sobolev-Schwartz distributions.

Chapter 2

Volterra Equations of the First Kind with Piecewise Continuous Kernels

2.1 Introduction. The State of the Art

The theory of integral models of evolving systems was initiated by L. Kantorovich, V. Glushkov and R. Solow in the mid-20th Century. Readers may refer to the bibliography in publications of [Hritonenko and Yatsenko (1996, 1999, 2003); Apartsyn (2003); Brunner (2004); Denisov and Lorenzi (1995); Markova *et al.* (2011); Boikov and Tynda (2003)] and the Introduction of this book. These works use VIEs of the first kind where bounds on the integration interval can be functions of time.

In this chapter we address the following integral equation:

$$\int_0^t K(t,s)x(s)ds = f(t), \ 0 < t \leq T, \tag{2.1.1}$$

where the kernel is defined as follows:

$$K(t,s) = \begin{cases} K_1(t,s), & 0 \leq s < \alpha_1(t), \\ K_2(t,s), & \alpha_1(t) \leq s < \alpha_2(t), \\ \cdots & \cdots \cdots \cdots \\ K_n(t,s), & \alpha_{n-1}(t) \leq s < t, \end{cases} \tag{2.1.2}$$

$$0 < \alpha_1(t) < \alpha_2(t) < \cdots < \alpha_{n-1}(t) < t, \ |\alpha_i'(t)| < 1.$$

The functions $K_i(t,s), i = \overline{1,n}, f(t)$ are continuous and sufficiently smooth, $f(0) = 0$.

It is to be noted that the conventional Glushkov integral model of evolving systems (here readers may refer to [Hritonenko and Yatsenko (1996); Markova *et al.* (2011)]) is the special case of the VIE (2.1.1) and (2.1.2) when all the kernels except $K_n(t,s)$ are zeros. For the theory and numerical solution of such equations $\int_{\alpha(t)}^t K(t,s)x(s)\,ds = f(t)$ the readers may refer to the monograph of [Apartsyn (2003)] and overview in the paper of [Brunner (2004)]. Excellent historical overview of the results concerning the VIEs of the first kind is given by [Brunner (1997)] in the paper *"1896–1996: One hundred years of Volterra integral equations of the first kind"*.

First results in the studies of the Volterra equations with discontinuous kernels were formulated by [Evans (1910)] in the beginning of 20th century. Results in the spectral theory of integral operators with discontinuous kernels were obtained by [Khromov (2006)]. Some results concerning the general approximation theory for integral equations with discontinuous kernels are presented by [Anselone (1967)]. For results of the treatment of Fredholm integral equation of the second kind with discontinuous kernels using product type quadrature formulas, readers may refer to paper of [Micke (1989)]. Asymptotic approximations of solutions of the Volterra equations of the first kind with analytical kernel $K(t, s)$ were constructed by [Magnitsky (1983)].

It is to be noted that solutions of the equation (2.1.1) can have an arbitrary constants and can be unbounded as $t \to 0$. For example, if

$$K(t, s) = \begin{cases} 1, \, 0 \leq s < t/2, \\ -1, \, t/2 \leq s \leq t, \end{cases} \qquad (2.1.3)$$

$f(t) = t$, then equation (2.1.1) has the solution $x(t) = c - \frac{\ln t}{\ln 2}$, where c is constant.

In this chapter we employ results of the papers of [Magnitsky (1983); Sidorov and Sidorov (2006); Sidorov and Trufanov (2009); Sidorov *et al.* (2007)] in order to formulate the algorithm for construction of the continuous solutions of equation (2.1.1) for $0 < t \leq T$ in the following form:

$$x(t) = \sum_{i=0}^{N} x_i(\ln t) t^i + t^N u(t). \qquad (2.1.4)$$

The coefficients $x_i(\ln t)$ are constructed as polynomials on powers of $\ln t$ and they may depend on certain number of arbitrary constants. N defines the necessary smoothness of the functions $K_i(t, s)$, $f(t)$. We describe an algorithm for construction of the function $u(t)$ in representation of the desired solution (2.1.4) based on successive approximations method which is uniformly converge on $[0, T]$. It is to be noted, that logarithmic-power asymptotic has been efficiently employed for solution of integral and differential equations in the singular cases, see results of [Magnitsky (1983)], [Sidorov and Sidorov (2011b)] and [Sidorov and Trufanov (2009)].

Outline

This chapter is organized as follows. In Section 2.2 after the problem statement we introduce the structure of solution and prove the existence theorem. The method for the asymptotic approximations construction is suggested in Section 2.4. The main theorem is formulated and proved also in Section 2.4. In Section 2.5 the illustrative example is presented. Finally, the concluding remarks are given.

2.2 The Structure of Solutions

For sake of clarity let us write $\alpha_i(t) = \alpha_i t$, where $0 = \alpha_0 < \alpha_1 < \cdots < \alpha_n = 1$. We first introduce the following condition:

A. Functions $K_i(t,s)$ are continuous and $\frac{\partial K_i(t,s)}{\partial t}$ are continuous and attain the finite limits at the boundaries of their ranges $D_i = \{t, s \mid \alpha_i t \leq s < \alpha_i t\}$. Functions $K_n(t,t) \neq 0$ for $t \in [0,T]$ and N are selected to satisfy the following inequality

$$\max_{0 \leq t \leq T} |K_n(t,t)|^{-1} \sum_{i=1}^{n} \left(\alpha_i^{1+N} |K_i(t, \alpha_i t)| + \alpha_{i-1}^{1+N} |K_i(t, \alpha_{i-1} t)| \right)$$

$$\leq 1 + q, \tag{2.2.1}$$

where $q < 1$, $\alpha_0 = 0, \alpha_n = 1$.

We note that condition **A** is satisfied for large enough N since $\alpha_i \in (0,1)$ for $i = \overline{1, n-1}$.

Lemma 2.1. *Let condition* **A** *be satisfied, let all the functions $K_i(t,s), i = \overline{1,n}$ be differentiable w.r.t. t and continuous w.r.t. s. Then the homogeneous equation*

$$\int_0^t K(t,s) s^N u(s) ds = 0 \tag{2.2.2}$$

possesses only the trivial solution in the space $C_{[0,T]}$.

Proof. Let us differentiate the equation (2.2.2) and take into account the jump discontinuity (2.1.2) of the kernel. Then we get an equivalent integral-functional equation

$$Lu + \sum_{i=1}^{n} \int_{\alpha_{i-1} t}^{\alpha_i t} \frac{K_i'(t,s)}{K_n(t,t)} \left(\frac{s}{t} \right)^N u(s) ds = 0, \tag{2.2.3}$$

where

$$Lu = \sum_{i=1}^{n} \left(\alpha_i^{1+N} K_i(t, \alpha_i t) u(\alpha_i t) \right.$$

$$\left. - \alpha_{i-1}^{1+N} K_i(t, \alpha_{i-1} t) u(\alpha_{i-1} t) \right) K_n^{-1}(t, t).$$

Due to the condition **A** in the space $\mathcal{C}_{[0,T]}$ we have the following bound

$$||Lu - u|| \leq q||u||.$$

Therefore, according to the Inverse Operator Theorem (here readers may refer to the monograph of [Trénoguine (1985)], p.134 and [Kreyszig (1978)]), and because of the inequalities $0 = \alpha_0 < \alpha_1 < \cdots < \alpha_n = 1$ there exists the following bounded inverse operator $L^{-1} \in \mathcal{L}(\mathcal{C}_{[0,T]} \to \mathcal{C}_{[0,T]})$ such that

$$||L^{-1}|| \leq \frac{1}{1-q} \qquad (2.2.4)$$

and equation (2.2.2) can be reduced as follows

$$u(t) = -L^{-1} \sum_{i=1}^{n} \int_{\alpha_{i-1}t}^{\alpha t} \frac{K_i'(t,s)}{K_n(t,t)} (s/t)^N u(s) ds \equiv (Au)(t), \qquad (2.2.5)$$

where $0 \leq t \leq T$. Let us introduce the following equivalent norm $||u|| = \max_{0 \leq t \leq T} e^{-lt}|u(t)|$, $l > 0$ in the space $\mathcal{C}_{[0,T]}$. In this norm the inequality (2.2.4) remains correct and for sufficiently large l A will be contraction operator since $||A|| \leq \frac{1}{1-q} m(l)$, where $m(l) \to 0$ as $l \to +\infty$. Therefore equation (2.2.5) possesses only the trivial solution. $\qquad \square$

Corollary 2.2.1. *Let all the condition of Lemma 2.1 be satified, and in addition $g(t) \in \mathcal{C}_{[0,T]}^{(1)}$, $|g'(t)| = o(t^N)$ as $t \to +0$. Then the inhomogeneous equation $\int_0^t K(t,s) s^N u(s) ds = g(t)$ possesses a unique solution, and $u(t) \to 0$ as $t \to +0$.*

The proof is trivial since differentiation of this equation leads to the equivalent equation

$$u(t) = Au + t^{-N} L^{-1} g'(t), \qquad (2.2.6)$$

where the right affine operator has linear part and $||A|| < 1$.

2.3   Existence Theorem

Theorem 2.1. *In the space of continuous functions on* $(0, T]$ *which have the finite limit for* $t \to +0$ *(briefly, in class* $\mathcal{C}_{(0,T]}$*), suppose there exists the function* $x^N(t)$ *such as for* $t \to +0$

$$\left(-\int_0^t K(t,s) x^N(s) ds + f(t) \right)' = o(t^N),$$

$f(t) \in \mathcal{C}^{(1)}_{[0,T]}$, $f(0) = 0$. *Then equation (2.1.1) possesses the following solution*

$$x(t) = x^N(t) + t^N u(t) \tag{2.3.1}$$

in class $\mathcal{C}_{(0,T]}$. *Here the function* $u(t) \in \mathcal{C}_{[0,T]}$, $u(t) \to 0$ *as* $t \to +0$ *is unique and it can be constructed by means of the successive approximations method.*

Proof. The proof follows from the Corollary 2.2.1. Indeed, with (2.3.1) we can rewrite the equation (2.1.1) as follows

$$\int_0^t K(t,s) s^N u(s) ds = g(t), \tag{2.3.2}$$

where

$$g(t) = -\int_0^t K(t,s) x^N(s) ds + f(t) \tag{2.3.3}$$

and satisfies the conditions in corollary. Therefore in (2.3.1) the function $u(t)$ can be uniquely constructed with successive approximations from the equation (2.2.6) using arbitrary initial condition. □

Definition 2.3.1. The equation (2.3.2) with the source part (2.3.3) we call the *regularization* of the equation (2.1.1), and the function $x^N(t)$ as asymptotic approximation of the solution (2.3.1) to the equation (2.1.1).

It is to be noted that one could numerically find the function $u(t)$ by solution of the equation (2.3.2) based on well-known numerical quadrature schemes (see the bibliography in the monographs of [Reinhardt (1985); Apartsyn (2003)]). The method of constructing asymptotic approximations $x^N(t)$ in the solution (2.3.2) will be studied below.

2.4   The Method of Asymptotic Approximations Construction

Let us suppose that along with the condition **A** the following condition **B** is satisfied.

B. Functions $K_i(t,s), i = \overline{1,n}$, $f(t)$ are $N+1$ times differentiable in the neighborhood of zero, where N is selected to satisfy the condition **A.**

We introduce an auxiliary algebraic equation w.r.t. $j \in \mathbb{N}$

$$L(j) \triangleq \sum_{i=1}^{n} K_i(0,0)(\alpha_i^{1+j} - \alpha_{i-1}^{1+j}) = 0 \qquad (2.4.1)$$

and name it as *characteristic equation* of the integral equation (2.1.1). Since $f(0) = 0$, then the equation

$$\sum_{i=1}^{n}(\alpha_i K_i(t,\alpha_i t)x(\alpha_i t)) - \alpha_{i-1}K_i(t,\alpha_{i-1}t)x(\alpha_{i-1}t)))$$

$$+ \sum_{i=1}^{n} \int_{\alpha_{i-1}t}^{\alpha_i t} K_i'(t,s)x(s)ds = f'(t)$$

is equivalent to the equation (2.1.1). We will look for the asymptotical approximation of its solution as a polynomial $x^N(t) = \sum_{j=0}^{N} x_j(\ln t)t^j$ with functional coefficients $x_j(\ln t)$. By means of the method of undetermined coefficients, and taking into account the inequalities $0 = \alpha_0 < \alpha_1 < \cdots < \alpha_n = 1$ we construct the recursive sequence of difference equations w.r.t. the coefficients $x_j(z)(z = \ln t)$ as follows:

$$K_n(0,0)x_j(z) + \sum_{i=1}^{n-1}\alpha_i^{1+j}\left(K_i(0,0)\right.$$

$$-K_{i+1}(0,0))x_j(z + a_i) = M_j(x_0,\ldots,x_{j-1}), \qquad (2.4.2)$$

where $j = \overline{0,N}$, $a_i = \ln \alpha_i$, $i = \overline{1, n-1}$, $M_0 = f'(0)$.

Here we follow the book of [Gelfond (1971)], p.330 and we seek the solution of the homogeneous difference equations in the form of $x(z) = \lambda^z$.

Substitution of the function λ^z into the homogeneous difference equations leads to $N+1$ equations for difference equations (2.4.2):

$$\mathcal{P}_j(\lambda) \equiv K_n(0,0) + \sum_{i=1}^{n-1} \alpha_i^{1+j}\left(K_i(0,0) - K_{i+1}(0,0)\right)\lambda^{a_i} = 0,$$

$$(2.4.3)$$

$$j = \overline{0, N}.$$

Therefore we have

Property. The jth equation (2.4.3) has the root $\lambda = 1$ if and only if j satisfies the characteristic equation (2.4.1) of the integral equation (2.1.1). Moreover, the multiplicity of the root j of the equation (2.4.1) is equal to r_j iff

$$L(j) = \sum_{i=1}^{n} K_i(0,0)(\alpha_i^{1+j} - \alpha_{i-1}^{1+j}) = 0, \qquad (2.4.4)$$

$$\sum_{i=1}^{n-1} \alpha_i^{1+j}\left(K_i(0,0) - K_{i+1}(0,0)\right)a_i^l = 0, l = \overline{1, r_j - 1}, \qquad (2.4.5)$$

$$\sum_{i=1}^{n-1} \alpha_i^{1+j}\left(K_i(0,0) - K_{i+1}(0,0)\right)a_i^{r_j} \neq 0,$$

where $\alpha_0 = 0, \alpha_n = 1, a_0 = 0, a_n = 0, a_i = \ln\alpha_i, i = \overline{1, n-1}$, and multiplicity $r_j \leq n - 1$.

This follows from the equality

$$\sum_{i=1}^{n-1} \alpha_i^{1+j} K_{i+1}(0,0) = \sum_{i=2}^{n} \alpha_{i-1}^{1+j} K_i(0,0) \qquad (2.4.6)$$

and from the structure of the equations (2.4.1)–(2.4.3).

If we suppose that for certain j multiplicity $r_j \geq n$, then $K_1(0,0) = K_2(0,0) = \cdots = K_{n-1}(0,0) = K_n(0,0)$ due to (2.4.5), since $\det \|a_i^l\|_{i,l=\overline{1,n}} \neq 0$. But due to (2.4.4) $K_n(0,0) = 0$, which contradicts **A**.

Under the conditions **A** and **B** there are two cases.

Regular Case

Let $L(j) \neq 0, j \in \mathbb{N}$. Then $\lambda = 1$ does not satisfy any of the equations in the sequence (2.4.3). All the coefficients x_i of the asymptotic expansion $x^N = \sum_{i=0}^{N} x_i t^i$ can be determined uniquely by means of the method of undetermined coefficients and do not depend upon $\ln t$.

Therefore we have the following theorem

Theorem 2.2. *Let the conditions* **A**, **B** *be satisfied and* $L(j) \neq 0, j \in \mathbb{N}$. *Then equation (2.1.1) possesses in* $C_{[0,T]}$ *the solution* $x(t) = \sum\limits_{i=0}^{N} x_i t^i + t^N u(t)$, *where* x_i *are determined uniquely by means of the method of undetermined coefficients and function* $u(t)$ *is uniquely constructed with successive approximations from equation (2.3.2).*

Singular Case

Let $L(j) = 0$ only for $j \in \{j_1, \ldots, j_k\} \subset \{0, 1, \ldots, N\}$ and multiplicity of the root $\lambda = 1$ (for the corresponding characteristic equation) is r_j. Let $M_j(z)$ in jth difference equation (2.4.2) be polynomial z of the order $n_j \geq 0$. Then in the singular case, i.e., for $r_j \geq 1$, based on Gelfond's method (readers may refer here to the book of [Gelfond (1971)]) the particular solution of the jth equation (2.4.2) we have to search as following polynomial $\hat{x}(z) = \sum\limits_{i=r_j}^{n_j+r_j} c_i z^i$.

Coefficients c_i of this polynomial can be sequentially calculated by means of the method of undetermined coefficients starting from $c_{n_j+r_j}$. Coefficient $x_j(z)$ of the desired asymptotical approximation x^N in this case is

$$x_j(z) = c_0 + c_1 z + \cdots + c_{r_j-1} + \hat{x}(z).$$

In singular case when $r_j \geq 1$, constants $c_0, \ldots, c_{r_j-1}$ remain arbitrary since the functions z^i, $i = 0, 1, \ldots, r_j - 1$ satisfy jth homogenous difference equation corresponding to (2.4.2).

In applications one could use the listed above Propety, and coefficient $x_j(z)$ in the singular case directly as a polynomial $\sum\limits_{i=0}^{n_j+r_j} c_i z^i$, where $c_{n_j+r_j}, \ldots, c_0$ are determined sequentially my means of the method of undetermined coefficients. $c_{r_j-1}, \ldots, c_0$ remains arbitrary. Therefore in the singular case when $L(j) = 0$ for some j new arbitrary constants r_j appear when the coefficient $x_j(z)$ will be determined.

Therefore we have the following Theorem:

Theorem 2.3. *Let conditions* **A** *and* **B** *be satisfied. Let the characteristic equation* $L(j) = 0$ *of the integral equation (2.1.1) has exactly* k *natural roots* $\{j_1, \ldots, j_k\}$. *And let the root* $\lambda = 1$ *of* jth *equation (2.4.3) has multiplicity*

r_j. Then equation (2.1.1) possesses the following solution in $\mathcal{C}_{(0,T]}$

$$x = \sum_{i=0}^{N} x_i(\ln t)t^i + t^N u(t),$$ (2.4.7)

which depends on $p = r_1 + \cdots + r_k$ arbitrary constants. Moreover, co-efficients x_i in the asymptotic approximation $x^N(t)$ are the polynomials depending on $\ln t$. The function $u(t)$ can be constructed by successive approximations which converge uniformly for $t \in [0, T]$.

It is to be noted that function $u(t)$ can constructed numerically from (2.3.2).

Remark 2.4.1. If $L(0) = 0$, then in (2.4.7) $x_0 = const + a \ln t$, where a is the defined constant. Therefore in this case $x(t) \in \mathcal{C}_{(0,T]}$, $\lim_{t \to +0} x(t) = \infty$.

Remark 2.4.2. These results can be generalized if in the equation (2.1.1) for $\alpha_{i-1}(t) \leq s \leq \alpha_i(t)$ $K(t, s) = K_i(t, s)$ where $\alpha_i(0) = 0$, $0 \leq \alpha_0'(0) < \alpha_1'(0) < \cdots < \alpha_n'(0) \leq 1$, $i = 1, \ldots, n$.

2.5   Example

Example 2.1. Consider the following VIE

$$\int_0^t K(t, s)x(s)ds = t, \quad 0 < t < \infty,$$ (2.5.1)

where the kernel $K(t, s)$ is defined by

$$K(t, s) = \begin{cases} 1 + t - 2s, & 0 \leq s < t/2, \\ -1, & t/2 \leq s \leq t, \end{cases}$$ (2.5.2)

The characteristic polynomial

$$L(j) \equiv \left(\frac{1}{2}\right)^{1+j} - \left(1 - \left(\frac{1}{2}\right)^{1+j}\right) = 0$$

has the simple unique root $j = 0$, therefore the solution to the VIE (2.5.1) must has $\ln t$ in the power of one only, i.e., the solution can be decomposed into the following series (which are converging series for $0 \leq t < \infty$)

$$\hat{x}(t) = \ln t \sum_{n=0}^{\infty} a_n t^n + \sum_{n=1}^{\infty} b_n t^n.$$

The method of undetermined coefficients gives: $a_0 = a_1 = -1/\ln 2$, $b_1 = 2 + 1/\ln 2$,

$$a_n = (-1)^n \frac{1}{n! \prod\limits_{k=2}^{n} (1 - 2^k) \ln 2}, \quad n = 2, 3, \ldots$$

$$b_n = \frac{1}{1 - 2^n} \left\{ a_n \ln 2 + a_{n-1} \frac{1}{n} \left(\ln 2 + \frac{1}{n+1} \right) - b_{n-1} \frac{1}{n} \right\}, n = 2, 3, \ldots.$$

The corresponding homogeneous integral equation has the solution

$$\phi(t) = 1 + \sum_{n=1}^{\infty} \frac{1}{n! \prod\limits_{k=1}^{n} (2^k - 1) \ln 2} t^n.$$

Therefore the VIEq (2.5.1) for $0 \leq t < \infty$ has the following solution $x(t) = c\phi(t) + \hat{x}(t)$.

Chapter 3

Volterra Matrix Equation of the First Kind with Piecewise Continuous Kernels

3.1 Introduction

In this chaper we address the following matrix VIE of the first kind

$$\int_0^t K(t,s)x(s)ds = f(t), \ 0 < t \le T. \tag{3.1.1}$$

We assume that the $m \times m$ matrix kernel $K(t,s)$ has jump discontinuities on curves $s = \alpha_i(t)$, $i = 1, \ldots, n-1$, $0 \le s \le t \le T$ as follows

$$K(t,s) = \begin{cases} K_1(t,s), & 0 \le s \le \alpha_1(t), \\ K_2(t,s), & \alpha_1(t) < s \le \alpha_2(t), \\ \cdots & \cdots\cdots \\ K_n(t,s), & \alpha_{n-1}(t) < s \le t, \end{cases} \tag{3.1.2}$$

$f(t) = (f_1(t), \ldots, f_m(t))'$, $x(t) = (x_1(t), \ldots, x_m(t))'$. $K_i(t,s)$ are $m \times m$ matrices. We assume that these matrices are defined, continuous and have continuous derivatives (w.r.t. t) in the sectors $D_i = \{s,t \,|\, \alpha_{i-1}(t) < s \le \alpha_i(t)\}$, $i = \overline{1,n}$, $\alpha_0 = 0$, $\alpha_n(t) = t$. The functions $f_i(t), \alpha_i(t)$ have continuous derivatives, $f_i(0) = 0$, $\alpha_i(0) = 0$, $0 < \alpha_1'(0) < \alpha_2'(0) < \cdots < \alpha_{n-1}'(0) < 1$, $0 < \alpha_1(t) < \alpha_2(t) < \cdots < \alpha_{n-1}(t) < t$ for $t \in (0,T]$. In the present chapter we search for the continuous solutions of the equation (3.1.1) for $t \in (0, T']$, where $0 < T' \le T$, and we note that $\lim_{t \to +0} x(t)$ can also be infinite. We assume below that each matrix $K_i(t,s)$, $i = \overline{1,n}$ has a continuously differentiable (w.r.t. t) extension on the interval $0 \le s \le t \le T$. Homogeneous systems can possess nontrivial solutions. Differentiation of the system (3.1.1) w.r.t. t in contrast to the classic cases addressed by [Reinhardt (1985); Asanov (1992); Hritonenko and Yatsenko (1996); Denisov and Lorenzi (1995); Apartsyn (2003)] lead us to a new class of the integral equations with functionally perturbed argument. That is the reason why the conventional techniques (discussed in the books of [Pruss (2012); Reinhardt (1985); Apartsyn (2003)]) are not applicable to studies of systems (3.1.1) with jump discontinuous kernels and they are of both theoretical and practical interest. Operator equations with functionally perturbed argument are addressed by [Sidorov and Trufanov (2009); Sidorov *et al.* (2007)].

Outline

This chapter is organized as follows. In Section 3.2 the algorithm for construction of the logarithmic-power asymptotic

$$\hat{x}(t) = \sum_{i=0}^{N} x_i(\ln t)t^i \qquad (3.1.3)$$

of the desired continuous solutions of the system (3.1.1) is presented. In particular regular and non-regular cases are identified based on the characteristic equation's roots and these are then studied. Next in Section 3.3 we prove the Existence Theorem of the parametric families of solutions of the system (3.1.1). Finally, in Section 3.4 we derive sufficient conditions for existence and uniqueness of continuous solution and present an illustrative example.

3.2   Construction of Asymptotic Approximation of Solutions to Inhomogeneous System

Let the following condition be satisfied.

A. There are matrices $\mathcal{P}_i = \sum\limits_{\nu+\mu=1}^{N} K_{i\nu\mu}t^\nu s^\mu$, $i = \overline{1,n}$, vector-function $f^N(t) = \sum\limits_{\nu=1}^{N} f_\nu t^\nu$, and polynomials $\alpha_i^N(t) = \sum\limits_{\nu=1}^{N} \alpha_{i\nu}t^\nu$, $i = \overline{1,n-1}$, where $\alpha_{i1} \in [0,1)$, such as for $t \to +0$, $s \to +0$ the following estimates are satisfied

$$||K_i(t,s) - \mathcal{P}_i(t,s)||_{\mathcal{L}(\mathbb{R}^n \to \mathbb{R}^n)} = \mathcal{O}((t+s)^{N+1}), \ i = \overline{1,n},$$

$$||f(t) - f^N(t)||_{\mathbb{R}^n} = \mathcal{O}(t^{N+1}),$$

$$|\alpha_i(t) - \alpha_i^N(t)| = \mathcal{O}(t^{N+1}), i = \overline{1,n-1}.$$

We call the expansions in $\mathbf{A}$ in powers of t and of s the Taylor polynomials. Let us introduce the matrix

$$B(j) = K_n(0,0) + \sum_{i=1}^{n-1} (\alpha_i'(0))^{1+j} (K_i(0,0) - K_{i+1}(0,0)) \qquad (3.2.1)$$

and the following algebraic equation

$$L(j) \stackrel{\text{def}}{=} \det B(j) = 0.$$

Similarly as in the previous chapter we call it the *characteristic equation* of the system of integral equations (3.1.1). Since $f(0) = 0$ the matrices $K_i(t,s)$ and vector $f(t)$ have continuous derivatives w.r.t. t, then differentiation of the system (3.1.1) lead us to the equivalent system of integral equations with functional perturbations of argument

$$F(x) \stackrel{\text{def}}{=} K_n(t,t)x(t) + \sum_{i=1}^{n-1} \alpha_i'(t) \{ K_i(t, \alpha_i(t))$$

$$-K_{i+1}(t, \alpha_i(t)) \} x(\alpha_i(t)) + \sum_{i=1}^{n} \int_{\alpha_{i-1}(t)}^{\alpha_i(t)} K_i^{(1)}(t,s)x(s)ds - f'(t) = 0,$$

$$(3.2.2)$$

where $\alpha_0 = 0$, $\alpha_n(t) = t$.

Remark 3.1. Since $\alpha_i(0) = 0$ we follow [Elsgoltz (1964)] and call equation (3.2.2) as *functional equation of neutral type*.

Here we do not assume that the homogeneous system of (3.1.1) has only the trivial solution. Hence the homogeneous integral-functional system corresponding to (3.2.2) can have nontrivial solutions. Here we follow the previous sections and seek an asymptotic approximation of a particular solution of the inhomogeneous equation (3.2.2) as the following polynomial

$$\hat{x}(t) = \sum_{j=0}^{N} x_j (\ln t) t^j. \qquad (3.2.3)$$

Let us demonstrate that the coefficients x_j depend on $\ln t$ and the free parameters in the general singular case. This is in line with possibility of the existence of nontrivial solutions of the homogeneous system.

The following regular and singular cases are possible.

Regular Case: $L(j) \neq 0$, $j \in (0, 1, \ldots, N)$

In this case the coefficients x_j are constant vectors from $\mathbb{R}^m$. Indeed, let us substitute the expansion (3.2.3) in the system (3.2.2) and apply the method of undetermined coefficients, taking into account the condition **A**. Then we get the recurrent sequence of linear systems of algebraic equations (SLAEs) with respect to the vectors x_j :

$$B(0)x_0 = f'(0), \qquad\qquad (3.2.4)$$

$$B(j)x_j = M_j(x_0, \ldots, x_{j-1}), \, j = 1, \ldots, N. \qquad (3.2.5)$$

The vector M_j can be expressed through the solutions $x_0, \ldots, x_{j-1}$ of the previous systems and coefficients of the Taylor polynomials from condition **A**.

Since in the regular case $\det B(j) \neq 0$, then the vectors $x_0, \ldots, x_N$ can be uniquely determined and the asymptotic (3.2.3) will therefore be constructed.

Non-regular case: equation $L(j) = 0$ has integer roots

Let us introduce the definitions:

Definition 3.1. We call the quantity j^* the *regular point of the matrix* $B(j)$ if the matrix $B(j^*)$ is invertible.

Definition 3.2. We call the quantity j^* a *simple singular point of a matrix* $B(j)$ if $\det B(j^*) = 0$, $\det\left[(B^{(1)}(j^*)\phi_i, \psi_l)\right]_{i,l=1}^{r} \neq 0$, where $\{\psi_i\}_1^r$ is a basis in $N(B'(j^*))$, $B'(j^*)$ is a transposed matrix, and $B^{(1)}(j)$ is the derivative of the matrix w.r.t. j.

Definition 3.3. We call the quantity j^* a *multiple singular point of the matrix* $B(j)$ (with multiplicity $k + 1$) if $\det B(j^*) = 0$, derivatives $B^{(1)}(j^*), \ldots, B^{(k)}(j^*)$ are zero matrices,

$$\det\left[(B^{(k+1)}(j^*)\phi_i, \psi_l)\right]_{i,l=1}^{r} \neq 0,$$

$k \geq 1$, $\{\phi_i\}_1^r$ is a basis in $N(B(j^*))$, $\{\psi_i\}_1^r$ is a basis in $N(B'(j^*))$.

We note that $B^{(k)}(j) = \sum\limits_{i=1}^{n-1} (\alpha'_i(0))^{1+j} a_i^k (K_i(0,0) - K_{i+1}(0,0))$, where $a_i = \ln \alpha'_i(0)$.

Remark 3.2. Let $m = 1$ (this case correspond to the single equation (3.1.1)). Then $B(j) = L(j)$. Hence in the case of the single equation the Definition 3.2 means that j is the single root of the characteristic equation $L(j) = 0$, and Definition 3.3 means that j is $(k+1)$ multiple root of this equation.

Let us demonstrate that in the singular case the coefficients x_j are polynomials in powers of $\ln t$ and depend on arbitrary constants. Order of polynomials and the number of arbitrary constants are associated with the multiplicity of the singular points of the matrices $B(j)$ and also the ranks of these matrices.

Indeed, since the coefficient x_0 in the singular case may depend on $\ln t$, we employ the method of undetermined coefficients and seek x_0 as the solution of the difference system

$$K_n(0,0)x_0(z) + \sum_{i=1}^{n-1} \alpha'_i(0)(K_i(0,0) - K_{i+1}(0,0))x_0(z + a_i) = f'(0), \quad (3.2.6)$$

where $a_i = \ln \alpha'(0), z = \ln t$.

Here there are three cases to consider:

1st Case.
$L(0) \neq 0$, i.e., $\det B(0) \neq 0$. Then the coefficient x_0 does not depend on z and is determined uniquely from the SLAE (3.2.5) with invertible matrix $B(0)$.

2nd Case.
Let $j = 0$ be the simple singular point of the matrix $B(j)$. We seek the coefficient $x_0(z)$ from the system (3.2.6) as a linear vector-function

$$x_0(z) = x_{01}z + x_{02}. \quad (3.2.7)$$

Let us substitute (3.2.7) in (3.2.6), and we get the following two SLAEs for determination of the vectors x_{01}, x_{02}:

$$B(0)x_{01} = 0, \quad (3.2.8)$$

$$B(0)x_{02} + B^{(1)}(0)x_{01} = f'(0). \quad (3.2.9)$$

Here $\det B(0) = 0$, $\{\phi_i\}_1^r$ is a basis in $N(B(0))$. Hence $x_{01} = \sum_{k=1}^{r} c_k \phi_k$. We substitute x_{01} into (3.2.9) and based on the Fredholm alternative, the vector $c = (c_1, \ldots, c_r)'$ can be determined uniquely from the following system of linear algebraic equations

$$\sum_{k=1}^{r}(B^{(1)}(0)\phi_k, \psi_i)c_k = (f'(0), \psi_i), \ i = \overline{1, r}$$

with non-singular matrix. The coefficient x_{02} can be determined from the system (3.2.9) with accuracy up to $\mathrm{span}(\phi_1, \ldots, \phi_r)$. Therefore, in the 2nd case the coefficient $x_0(z)$ will depend on r arbitrary constants and be linear w.r.t. z.

3rd Case.   Let $j = 0$ be the singular point of the matrix $B(j)$ of multiplicity $k + 1$ (see Definition 3.3), where $k \geq 1$. We construct solution to the difference system (3.2.6) as following polynomial

$$x_0(z) = x_{01}z^{k+1} + x_{02}z^k + \cdots + x_{0k+1}z + x_{0k+2}. \tag{3.2.10}$$

Let us substitute the polynomial (3.2.10) into the system (3.2.6). We next take into account the equality

$$\frac{d^k}{dj^k}B(j) = \sum_{i=1}^{n-1}(\alpha_i'(0))^{1+j}a_i^k(K_i(0,0) - K_{i+1}(0,0)),$$

where $a_i = \ln \alpha_i'(0)$ and equation coefficients of powers $z^{k+1}, z^k, \ldots, z, z^0$ to zero yields the recursive sequence of linear algebraic equation with respect to the coefficients $x_{01}, x_{02}, \ldots, x_{0k+2}$:

$$B(0)x_{01} = 0,$$

$$B(0)x_{02} + B^{(1)}(0)\binom{k+1}{k}x_{01} = 0,$$

$$B(0)x_{0l+1} + B^{(l)}(0)\binom{k+1}{k+1-l}x_{01} + B^{(l-1)}(0)\binom{k}{k+1-l}x_{02}$$

$$+ \cdots + B^{(1)}(0)\binom{k+1-l+1}{k+1-l}x_{ol} = 0, \ l = 1, \ldots, k,$$

$$B(0)x_{0k+2} + B^{(k+1)}(0)x_{01} + B^{(k)}x_{02} + \ldots B^{(1)}(0)x_{0k+1} = f'(0). \tag{3.2.11}$$

Since in this case due to the conditions on the derivatives $\left.\frac{d^i B(j)}{dj^i}\right|_{j=0}$ are zero matrices, then

$$x_{0i} = \sum_{j=1}^{r}c_{ij}\phi_j, i = 1, \ldots, k+1.$$

As result the system (3.2.11) can be rewritten as

$$B(0)x_{0,k+2} + B^{(k+1)}(0)x_{01} = f'(0). \tag{3.2.12}$$

Since $\det\left[(B^{(k+1)}(0)\phi_i, \psi_k)\right]_{i,k=\overline{1,r}} \neq 0$, then the vector $c^1 \overset{\text{def}}{=} (c_{11}, \ldots, c_{1r})'$
can be uniquly determined from the conditions of resolvability of the system
(3.2.12). Therefore,

$$x_{0\,k+2} = \sum_{j=1}^{r} c_{k+2\,j}\phi_j + \hat{x}_{k+2},$$

$\hat{x}_{k+2}$ is the particular solution to the SLAE (3.2.12). Vector $c^{k+2} \overset{\text{def}}{=}$
$(c_{k+2,1}, \ldots c_{k+2,r})'$, and vectors $c^2, \ldots, c^{k+1}$, remain arbitrary. Therefore
in the 3rd case the coefficient $x_0(z)$ is a polynomial of degree $k+1$ w.r.t.
z and it depends on $r \times (k+1)$ arbitrary constants. One may apply the
method of undetermined coefficients and take into account the following
equality

$$\int t^j \ln^k t\, dt = t^{j+1} \sum_{s=0}^{k} (-1)^s \frac{k(k-1)\ldots(k-(s-1))}{(j+1)^{s+1}} \ln^{k-s} t,$$

and construct the system of difference equations for determination of the
coefficient $x_1(z)$ $(z = \ln t)$. Indeed,

$$F(x)\Big|_{x=x_0+x_1 t} \overset{\text{def}}{=} \left[K_n(0,0)x_1(z) + \sum_{i=1}^{n-1} (\alpha_i'(0))^2 (K_i(0,0) \right.$$

$$\left. - K_{i+1}(0,0))x_1(z+a_i) + P_1(x_0(z)) \right] t + r(t), \quad r(t) = o(t).$$

$$\tag{3.2.13}$$

Here $P_1(x_0(z))$ is a certain polynomial on z, which degree is equal to the
multiplicity of the singular point $j = 0$ of the matrix $B(j)$. From (3.2.13)
due to the estimate $r(t) = o(t)$ as $t \to 0$ it follows that the coefficient $x_1(z)$
must satisfy the following system of difference equations

$$K_n(0,0)x_1(z) + \sum_{i=1}^{n-1} (\alpha'(0))^2 (K_i(0,0) - K_{i+1}(0,0))x_1(z+a_i)$$

$$+ P_1(x_0(z)) = 0. \tag{3.2.14}$$

If $j = 1$ is a regular point of the matrix $B(j)$, then system (3.2.14) possesses
a solution $x_1(z)$ as polynomial of the same order as multiplicity of the
singular point $j = 0$ of the matrix $B(0)$. If $j = 1$ is a singular point of the

matrix $B(j)$, then the solution $x_1(z)$ can be constructed as a polynomial of power $k_0 + k_1$, where k_0 and k_1 are multiplicities of singular points $j = 0$ and $j = 1$ of the matrix $B(j)$. The coefficient $x_1(z)$ depends on $r_0 k_0 + r_1 k_1$ arbitrary constants, where $r_0 = \dim N(B(0))$, $r_1 = \dim N(B(1))$.

Let us impose the following condition

B. Let the matrix $B(j)$ have only the regular points in the array $(0, 1, \ldots, N)$ or has singular points of multiplicities k_j.

Then, in a similar way we can calculate the remaining coefficients $x_2(z), \ldots, x_N(z)$ of $\hat{x}(t)$ from the sequence of difference equations

$$K_n(0,0)x_j(z) + \sum_{i=1}^{n-1} (\alpha'(0))^{1+j} \left(K_i(0,0) - K_{i+1}(0,0) \right) x_j(z + a_i)$$

$$+ \mathcal{P}_j(x_0(z), \ldots, x_{j-1}(z))) = 0, \; j = \overline{2, N}.$$

Hence we have the following

Lemma 3.1. *Let conditions* **A** *and* **B** *be satisfied. Then exists vector function* $\hat{x}(t) = \sum_{i=0}^{N} x_i(\ln t) t^i$, *such as when* $t \to +0$ *the following estimate is satisfied* $|F(\hat{x}(t))|_{\mathbb{R}^m} = o(t^N)$ *and the coefficients* $x_i(\ln t)$ *are polynomials of* $\ln t$ *with increasing powers smaller then sum* $\sum_j k_j$ *of the singular points* j *of the matrix* $B(j)$ *from the array* $(0, 1, \ldots, i)$. *Th coefficients* $x_i(\ln t)$ *are determined up to* $\sum_{j=0}^{i} \dim N(B(j))k_j$ *arbitrary constants.*

3.3 Existence of Continuous Parametric Family of Solutions

Since $0 \le \alpha_i'(0) < 1$, $\alpha_i(0) = 0$, $i = \overline{1, n-1}$, then for any $0 < \varepsilon < 1$ exists $T' \in (0, T]$ such as $\max\limits_{i=\overline{1,n-1}, t \in [0,T']} |\alpha_i'(t)| \le \varepsilon$ and $\sup\limits_{i=\overline{1,n-1}, t \in (0,T']} \frac{\alpha_i(t)}{t} \le \varepsilon$.

Let the following condition be satisfied

C. Let $\det K_n(t, t) \neq 0$, $t \in [0, T']$ and N^* is selected large enough to have the following inequality satisfied

$$\max_{t \in [0,T]} \varepsilon^{N^*} |K_n^{-1}(t,t)|_{\mathcal{L}(\mathbb{R}^m \to \mathbb{R}^m)}$$

$$\times \sum_{i=1}^{n-1} |\alpha_i^{(1)}(t)| |K_i(t, \alpha_i(t)) - K_{i+1}(t, \alpha_i(t))|_{\mathcal{L}(\mathbb{R}^m \to \mathbb{R}^m)} \le q < 1,$$

$$(3.3.1)$$

where $|\cdot|_{\mathcal{L}(\mathbb{R}^m \to \mathbb{R}^m)}$ is norm of the $m \times m$ matrix.

Lemma 3.2. *Let the condition* **C** *be satisfied. Let* $C_{(0,T']}$ *be the class of vector functions continuous for* $t \in (0, T']$ *and having a limit (possibly infinite) in* $t \to +0$. *Let in* $C_{(0,T']}$ *exists the element* $\hat{x}(t)$ *such as the following estimate*

$$|F(\hat{x}(t))|_{\mathbb{R}^m} = o(t^N), N \geq N^*$$

is satisfied. Then equation (3.2.2) in $C_{(0,T']}$ *has the solution*

$$x(t) = \hat{x}(t) + t^{N^*} u(t), \tag{3.3.2}$$

where $u(t) \in C_{[0,T']}$ *and the solution can be uniquely defined with successive approximations.*

Proof. Let us substitute (3.3.2) in the equation (3.2.2). We get the following integral-functional system for determination of $u(t)$

$$K_n(t,t)u(t) + \sum_{i=1}^{n-1} \alpha_i'(t) \left(\frac{\alpha_i(t)}{t}\right)^{N^*} \left(K_i(t, \alpha_i(t))\right.$$

$$\left. -K_{i+1}(t, \alpha_i(t))\right) u(\alpha_i(t))$$

$$+ \sum_{i=1}^{n} \int_{\alpha_{i-1}(t)}^{\alpha_i(t)} K_i^{(1)}(t,s) \left(\frac{s}{t}\right)^{N^*} u(s)ds + F(\hat{x}(t))/t^{N^*} = 0. \tag{3.3.3}$$

Let us introduce the linear operators:

$$Lu \overset{\text{def}}{=} K_n^{-1}(t,t) \sum_{i=1}^{n-1} \alpha_i'(t) \left(\frac{\alpha_i(t)}{t}\right)^{N^*} \left\{K_i(t, \alpha_i(t))\right.$$

$$\left. -K_{i+1}(t, \alpha_i(t))\right\} u(\alpha_i(t)),$$

$$Ku \overset{\text{def}}{=} \sum_{i=1}^{n} \int_{\alpha_{i-1}(t)}^{\alpha_i(t)} K_n^{-1}(t,t)K_i^{(1)}(t,s)(s/t)^{N^*} u(s)ds.$$

Then we can present the system (3.3.3) as follows

$$u + (L + K)u = \gamma(t),$$

where $\gamma(t) = K_n^{-1}(t,t)F(x^N(t))/t^{N^*}$ is continuous vector function. Let us introduce the Banach space X of continuous vector-functions $u(t)$ with norm

$$||u||_l = \max_{0 \le t \le T'} e^{-lt}|u(t)|_{\mathbb{R}^m}, \, l > 0.$$

Hence because of inequalities $\sup\limits_{t \in [0,T]} \frac{\alpha_i(t)}{t} \le \varepsilon < 1$ and condition **C** for any $l \ge 0$ norm of linear functional operator L attains the following bound

$$||L||_{\mathcal{L}(X \to X)} \le q < 1.$$

Moreover for the integral operator K for large enough l the following bound is satisfied

$$||K||_{\mathcal{L}(X \to X)} \le q_1 < 1 - q.$$

As a result for large enough $l > 0$ we have

$$||L + K||_{\mathcal{L}(X \to X)} < 1,$$

i.e., the linear operator $L + K$ is a contractive operator in X. Hence the sequence $u_n = -(L + K)u_{n-1} + \gamma(t)$, $u_0 = \gamma(t)$ converges.

$\square$

Theorem 3.1 (The Main Theorem). *Let the following conditions **A**, **B** and **C**, $f(0) = 0$ be satisfied. Let also the matrix $B(j)$ has exactly ν singular points $j_1, \ldots, j_\nu$ of multiplicities k_i, $i = \overline{1, \nu}$, in the array $(0, 1, \ldots, N)$ and the rest of the points of the array are regular. Let rank $B(j_i) = r_i$, $i = \overline{1, \nu}$.*

Then equation (3.1.1) has the solution

$$x(t) = \hat{x}(t) + t^{N^*} u(t),$$

depending on $\sum\limits_{i=1}^{\nu} (m - r_i)k_i$ arbitrary constants, $0 < t \le T' \le T$.

Proof. Due to the conditions of the Lemma 3.1 and imposed conditions of Theorem 3.1 it makes possible the construction of asymptotic approximation $\hat{x}(t)$ of the desired solution in the form of logarithmic-power polynomial $\sum\limits_{i=0}^{N} x_i(\ln t)t^i$. Moreover the coefficients $x_i(\ln t)$ will depend on the specified number of arbitrary constants. Because of the Lemma 3.2 we can apply the substitution $x(t) = \hat{x}(t) + t^{N^*} u(t)$, and continuous function $u(t)$ can be constructed with successive approximations. $\square$

Remark 3.3. In the Main Theorem's conditions for the asymptotic approximation $\hat{x}(t)$ of the desired solution the following asymptotic estimate $|x(t) - \hat{x}(t)|_{\mathbb{R}^n} = \mathcal{O}(t^{N^*})$, $t \to +0$ is satisfied.

Corollary 3.1. *Let* $\alpha_i(t) = \alpha_i t$, $i = \overline{1, n-1}$, $0 < \alpha_1 < \alpha_2 < \ldots < \alpha_{n-1} < 1$, *elements of the matrices* $K_i(t, s)$, $i = \overline{1, n-1}$ *and vector function* $f(t)$ *has an analytic extension to* $|s| < T$, $|t| < T$, $f(0) = 0$. *Matrix* $K_n(t, t)^{-1}$ *is analytical for* $|t| < T$. *Let* $\det B(j) \neq 0$, $j \in \mathbb{N} \cup 0$. *Then equation* (3.1.1) *possesses a unique solution* $x(t) = \sum_{i=0}^{\infty} x_i t^i$ *for* $0 \leq t < T$.

In some cases, a parametric family of solutions can be constructed in closed form.

Example 3.1. Let us consider the system

$$\int_0^{t/2} Kx(s)ds + \int_{t/2}^{t} (K - 2E)x(s)ds = dt,$$

$0 < t < \infty$, where K is symmetric constant matrix $m \times m$, $d \in \mathbb{R}^m$, $x(t) = (x_1(t), \ldots, x_m(t))'$, 1 is eigenvalue of the matrix K of rank r, $\{\phi_1, \ldots, \phi_r\}$ is corresponding orthonormal system of eigenvectors. This system possesses the parametric family of solutions

$$x(t) = -\ln t \sum_{i=1}^{r} \frac{(d, \phi_i)}{\ln 2} \phi_i + c_1 \phi_1 + \cdots + c_r \phi_r + \hat{a}.$$

Here $c_1, \ldots, c_r$ are arbitrary constants, vector $\hat{a}$ satisfies the SLAE

$$(K - E)\hat{a} = d - \sum_{i=1}^{r} (d, \phi_i)\phi_i.$$

3.4 Sufficient Conditions for Existence of Unique Continuous Solution

For sake of clarity in this section let us suppose $\alpha_i(t) = \alpha_i t$, $i = \overline{1, n-1}$, $0 < \alpha_1 < \alpha_2 < \cdots < \alpha_{n-1} < 1$. Let us introduce the matrix

$$D(t) \equiv \sum_{i=1}^{n-1} \alpha_i |K_n^{-1}(t,t)|_{\mathcal{L}(\mathbb{R}^m \to \mathbb{R}^m)} |(K_i(t, \alpha_i t) - K_{i+1}(t, \alpha_i t))|_{\mathcal{L}(\mathbb{R}^m \to \mathbb{R}^m)}.$$

Let the following condition be satisfied:

S. $D(0) < 1$; $\displaystyle\sup_{0 \le s \le t \le T} |K_n^{-1}(t,t)K(t,s)|_{\mathcal{L}(\mathbb{R}^m \to \mathbb{R}^m)} \le c < \infty.$

Here and below the matrix $K(t,s)$ is defined with formula (3.1.2).

Theorem 3.2. *(Sufficient conditions for the existence and uniqueness of solutions) Let the conditions* **S** *be satisfied, all the matrices* $K_i(t,s)$ *are continuous in* (3.1.2) *and they are continuously differentiable w.r.t. t, the vector $f(t)$ has continuous derivative, $f(0) = 0$. Then equation* (3.1.1) *possesses a unique solution in* $\mathcal{C}_{[0,T]}$. *Moreover, the solution can be found by the method of steps, combined with the method of successive approximations.*

Proof. Let us rewrite the equation (3.2.2) which is equivalent to the equation (3.1.1) as follows

$$x(t) + Ax + Kx = \overline{f}(t), \tag{3.4.1}$$

where the following notations are introduced

$$Ax \overset{\text{def}}{=} K_n^{-1}(t,t) \sum_{i=1}^{n-1} \alpha_i (K_i(t,\alpha_i t) - K_{i+1}(t,\alpha_i t)) x(\alpha_i t),$$

$$Kx \overset{\text{def}}{=} \sum_{i=1}^{n} \int_{\alpha_{i-1}t}^{\alpha_i t} K_n^{-1}(t,t) K_t^{(1)}(t,s) x(s)\,ds, \quad \overline{f}(t) = K_n^{-1}(t,t) f^{(1)}(t).$$

Let us fix $q < 1$ and select $h_1 > 0$ such as

$$\max_{0 \le t \le h_1} |D(t)|_{\mathcal{L}(\mathbb{R}^m \to \mathbb{R}^m)} = q < 1.$$

Because of the condition **S** exists such $h_1 > 0$. Let $0 < h < \min\{h_1, \frac{1-q}{c}\}$, where the constant c is defined in the condition **S**. We divide the interval $[0,T]$ into the intervals

$$[0,h],\ [h, h+\varepsilon h],\ [h+\varepsilon h, h+2\varepsilon h],\ldots \tag{3.4.2}$$

where ε is selected from $(0,1]$ such as $\alpha_{n-1} \le \frac{1}{1+\varepsilon}$. Denote by $x_0(t)$ the restriction of the desired solution $x(t)$ on the interval $[0,h]$ and let $x_n(t)$ be the restrictions to intervals

$$I_n = [(1 + (n-1)\varepsilon)h, (1 + n\varepsilon)h], n = 1, 2, \ldots.$$

Because of the selection of ε for $t \in I_n$ the "perturbed" argument $\alpha_i t \in \bigcup_{k=1}^{n-1} I_k$. Such an inclusion makes it possible to apply the known method of steps (here readers may refer to the books of [Smith (2010)] and [Elsgoltz (1964)].

For construction of the element $x_0(t) \in \mathcal{C}_{[0,h]}$ we construct the sequence $\{x_0^n(t)\}$:

$$x_0^n(t) = -Ax_0^{n-1} - Kx_0^{n-1} + \overline{f}(t),$$

$$x_0^0(t) = \overline{f}(t), \ t \in [0, h].$$

Because of the selection of h we have $||A + K||_{\mathcal{L}(\mathcal{C}_{[0,h]} \to \mathcal{C}_{[0,h]})} < 1$.

Hence the unique solution $x_0(t)$ of the equation (3.4.1) exists for $t \in [0, h]$. The sequence $x_0^n(t)$ converges uniformly to the unique solution $x_0(t)$. Let us continue the process for the desired solution construction for $t \geq h$, i.e., on the intervals I_n, $n = 1, 2, \dots$. To be specific let $\varepsilon = 1$ in (3.4.2).

Once we get $x_0(t) \in \mathcal{C}_{[0,h]}$ computed we construct $x_1(t)$ in the space $\mathcal{C}_{[h,2h]}$. We find $x_1(t)$ from the VIE of the 2nd kind

$$x(t) + \int_h^t K_n^{-1}(t,t) K_t'(t,s) x(s) ds = \overline{f}(t) - A x_0 - \int_0^h K_n^{-1}(t,t) K_t'(t,s) x_0(s) ds$$

using successive approximations. In this case $x_0(h) = x_1(h)$.

Let us introduce the continuous function

$$\overline{x}_1(t) = \begin{cases} x_0(t), & 0 \leq t \leq h, \\ x_1(t), & h \leq t \leq 2h, \end{cases} \tag{3.4.3}$$

which is the restriction of the desired continuous solution $x(t)$ on the interval $[0, 2h]$. Then the element $x_2(t) \in \mathcal{C}_{[2h,3h]}$ can be computed with successive approximations from the VIE of the second kind

$$x(t) + \int_{2h}^t K_n^{-1}(t,t) K_t'(t,s) x(s) ds = \overline{f}(t) - A \overline{x}_1 - \int_0^{2h} K_n^{-1}(t,t) K_t'(t,s) \overline{x}_1(s) ds.$$

Finally we construct the desired solution $x(t) \in \mathcal{C}_{[0,T]}$ to the equation (3.1.1) within N steps $(N \geq \frac{T}{h})$. □

Example 3.2. The integral equation

$$\int_0^{t/2} K_1(t-s) x(s) ds + \int_{t/2}^t K_2(t-s) x(s) ds = f(t), \ 0 < t \leq T,$$

where $K_1(t-s) = K_2(t-s) + E$ are matrices $m \times m$, E is unit matrix, $|K_2^{-1}(0)|_{\mathcal{L}(\mathbb{R}^m \to \mathbb{R}^m)} < 2$, matrix $K_2(t)$ and vector function $f(t)$ have continuous derivatives w.r.t. t, $f(0) = 0$, satisfies Theorem 3.2 conditions and has the unique continuous solution.

3.4.1 *Final Remarks on Volterra Matrix Equations*

In this chapter we methodologically continued the theory presented in Chapter 2 and studied the matrix Volterra integral equations of the first kind with jump discontinuous kernels, see also the article of [Sidorov

(2013)]. Then we present the algorithm for construction of the logarithmic-power asymptotic of the desired continuous solutions. Regular and non-regular cases are studied taking into account the characteristic equation's roots. The existence and uniqueness theorem is proved. We also derived the sufficient conditions for existence and uniqueness of continuous solution.

Chapter 4

Volterra Operator Equations of the First Kind with Piecewise Continuous Kernels

4.1 Introduction. The State of the Art

Let us introduce a domain $D = \{s, t; 0 < s < t < T\}$ in the plane s, t and define continuous functions $s = \alpha_i(t), i = \overline{1, n}$, which have continuous derivatives for $t \in (0, T)$. Let us suppose that $\alpha_i(0) = 0$, $0 < \alpha_1(t) < \cdots < \alpha_{n-1}(t) < t$ for $t \in (0, T)$, $0 < \alpha'_1(0) < \cdots < \alpha'_{n-1}(0) < 1$, and functions $s = \alpha_i(t)$, $i = \overline{0, n}$, $\alpha_0(t) = 0$, $\alpha_n(t) = t$, we split the domain D into disjoint sets: $D_1 = \{s, t : 0 \leq s < \alpha_1(t)\}$, $D_i = \{s, t : \alpha_{i-1}(t) < s < \alpha_i(t), i = \overline{2, n}\}$, such that $\overline{D} = \bigcup_1^n \overline{D_i}$. Let us introduce bi-parametric family of linear continuous operator-functions $K_i(t, s)$, defined for $t, s \in \overline{D_i}$, $i = \overline{1, n}$, which are differentiable w.r.t. t and acting from a Banach space E_1 into a Banach space E_2. Therefore $K_i(t, s) \in \mathcal{L}(E_1 \to E_2)$ and $\frac{\partial K_i(t, s)}{\partial t} \in \mathcal{L}(E_1 \to E_2)$ for $t, s \in \overline{D_i}, i = \overline{1, n}$. Let the space of continuous functions $x(t)$, defined on $[0, T]$ with ranges on E_1, be denoted by $\mathcal{C}_{([0,T];E_1)}$. We introduce the integral operator

$$\int_0^t K(t, s)u(s)ds \overset{\text{def}}{=} \sum_{i=1}^n \int_{\alpha_{i-1}(t)}^{\alpha_i(t)} K_i(t, s)x(s)ds \qquad (4.1.1)$$

with a piecewise continuous (or jump discontinuous) kernel

$$K(t, s) = \begin{cases} K_1(t, s), & t, s \in D_1, \\ \cdots & \cdots \cdots \\ K_n(t, s), & t, s \in D_n. \end{cases} \qquad (4.1.2)$$

In this text we concentrate on the equation

$$\int_0^t K(t, s)x(s)\, ds = f(t), \qquad (4.1.3)$$

where $f(t)$ and $f'(t)$ are functions with ranges in E_2 are defined and continuous for $t \in [0, T]$, $f(0) = 0$. We call the equation (4.1.3) the Volterra operator equation of the first kind. The objective of this chapter is to construct the solution of (4.1.3) in $C_{([0,T];E_1)}$. This problem and its numeric solution for $E_1 = E_2 = \mathbb{R}^1$, $n = 1$ has been studied in previous chapters.

Article of [Lorenzi (2013)] also deals with linear operator integral equations of the first kind and first-order in time integro-differential equations of degenerate type in a Banach space. The kernels are assumed to be piecewise continuous with linear closed operator values in X. Applications are given to linear integro-differential equations with kernels of "elliptic" and "parabolic" type.

It is to be noted that differentiation of the equation (4.1.3) leads to the integral-functional equation and, in general, its solution is not unique. The reader may also refer to the article of [Sidorov (2011b)]. Therefore construction of the solution to equation (4.1.3) cannot be carried out with classic analytical methods of the Volterra equations. In this chapter we address the equation (4.1.3) in the general case using the theory of operator equations with functionally perturbed argument of neutral type, here readers may refer to papers of [Sidorov and Trufanov (2009)], [Sidorov *et al.* (2007)], and to the textbook of [Elsgoltz (1964)].

Outline

This chapter is organized as follows. In Section 4.2 we continue and generalize our studies on the Volterra equations presented in the previous two sections. Readers may also refer to the publictions of [Sidorov and Sidorov (2006); Sidorov and Trufanov (2009); Sidorov *et al.* (2007, 2006, 2010c); Sidorov and Sidorov (2012); Sidorov (2011a); Sidorov and Sidorov (2011b)] for more details. and provide the sufficient conditions of existence and uniqueness of continuous solution of the equation (4.1.3) with piecewise continuous kernel (4.1.2). We also present two practical examples in order to illustrate the formulated existence and uniqueness theorem in Section 4.2. In Section 4.3 we construct the so called characteristic operator for the equation (4.1.3) and distinguish regular and non-regular cases based on the characteristic operator, in Section 4.4. Finally in Section 4.5 we give concluding remarks on Volterra operator equation of the first kind. It is to be noted here that the Volterra model of the evolving dynamical systems (for more details regarding evolving dynamical systems, readers may refer to [Apartsyn (2003)], and other monographs of [Hritonenko and Yatsenko (1996); Markova *et al.* (2011)]) is the special case of the abstract Volterra equations discussed in this chapter.

Here we suppose that an operator $K_n(t,t)$ has a bounded inverse for $t \in [0,T]$. Norms $||K_i(t,s)||_{\mathcal{L}(E_1 \to E_2)}$, $||\frac{\partial K_i(t,s)}{\partial t}||_{\mathcal{L}(E_1 \to E_2)}$, $i = \overline{1,n}$ of the linear operators are defined and they are continuous functions for $t, s \in \overline{D}_i$, $t \in [0,T]$.

4.2 Sufficient Conditions for Existence of the Unique Continuous Solution

Since $f(0) = 0$ differentiation of equation (4.1.3) leads to the equivalent functional-operator equation

$$F(x) \overset{\text{def}}{=} K_n(t,t)x(t) + \sum_{i=1}^{n-1} \alpha_i'(t)\{K_i(t,\alpha_i(t)) - K_{i+1}(t,\alpha_i(t))\}x(\alpha_i(t))$$

$$+ \sum_{i=1}^{n} \int_{\alpha_{i-1}(t)}^{\alpha_i(t)} K_i^{(1)}(t,s)x(s)\,ds - f'(t) = 0, \tag{4.2.1}$$

where $\alpha_0 = 0$, $\alpha_n(t) = t$.

Let us introduce the function

$$D(t) \overset{\text{def}}{=} \sum_{i=1}^{n-1} |\alpha_i'(t)| \, ||K_n^{-1}(t,t)\{K_i(t,\alpha_i(t)) - K_{i+1}(t,\alpha_i(t))\}||_{\mathcal{L}(E_1 \to E_1)},$$

where $|| \cdot ||_{\mathcal{L}(E_1 \to E_1)}$ is linear operator's norm. Let the following condition be satisfied:

A. $D(0) < 1$, $\displaystyle\sup_{0 < s < t < T} ||K_n^{-1}(t,t)K(t,s)||_{\mathcal{L}(E_1 \to E_1)} \leq c < \infty.$

Clearly the inequality $D(0) < 1$ is satisfied if $|\alpha_i^{(1)}(0)|$ are sufficiently small. Here and below an operator $K(t,s)$ is defined with formula (4.1.2) in $\bigcup_1^n D_i$. It's derivative w.r.t. t is defined as

$$K^{(1)}(t,s) = \begin{cases} K_1^{(1)}(t,s), & t,s \in D_1, \\ \dots \quad \dots\dots \\ K_n^{(1)}(t,s), & t,s \in D_n. \end{cases} \tag{4.2.2}$$

for $t, s \in \bigcup_1^n D_i$.

Theorem 4.1. *(Sufficient conditions for existence and uniqueness of the solution) Let the condition* **A** *be satisfied, all the operators* $K_i(t, s)$ *in (4.1.2) are continuous, and w.r.t. t they also have continuous derivatives, source $f(t)$ has continuous derivatives, $f(0) = 0$. Then equation (4.1.3) possesses the unique solution in the class of continuous functions $\mathcal{C}_{([0,T];E_1)}$. Moreover the solution can be constructed using the step method combined with successive approximations.*

Proof. Let us apply the operator $K_n^{-1}(t, t)$ to the equation (4.2.1). We get the following equation

$$x(t) + Ax + Kx = \overline{f}(t), \qquad (4.2.3)$$

where the following notations are introduced

$$A(t)x \stackrel{\text{def}}{=} K_n^{-1}(t, t) \sum_{i=1}^{n-1} \alpha_i^{(1)}(t)(K_i(t, \alpha_i(t)) - K_{i+1}(t, \alpha_i(t)))x(\alpha_i t),$$

$$Kx \stackrel{\text{def}}{=} \sum_{i=1}^{n} \int_{\alpha_{i-1}(t)}^{\alpha_i(t)} K_n^{-1}(t, t) K_i^{(1)}(t, s)x(s)ds,$$

$$\overline{f}(t) \stackrel{\text{def}}{=} K_n^{-1}(t, t)f^{(1)}(t).$$

Let us fix $q < 1$ and select $h_1 > 0$ such that $\max_{0 \leq t \leq h_1} D(t) = q < 1$. Such $h_1 > 0$ can be found because of condition **A**, and since the operator-functions $K_i(t, s)$ are continuous in the operator topology and because functions $\alpha_i(t)$, $\alpha_i'(t)$ are continuous. Let $0 < h < \min\{h_1, \frac{1-q}{c}\}$, where the constant c is defined in the condition **A**. Let us split $[0, T]$ into

$$[0, h], [h, h + \varepsilon h], [h + \varepsilon h, h + 2\varepsilon h], \ldots \qquad (4.2.4)$$

The contraction of the desired solution $x(t)$ onto the interval $[0, h]$ we denote as $x_0(t)$, the contraction onto

$$I_m = [(1 + (m-1)\varepsilon)h, (1 + m\varepsilon)h], \ m = 1, 2, \ldots$$

we denote as $x_m(t)$. Let us select ε from $(0, 1]$ such as for $t \in I_m$ the "perturbed" arguments $\alpha_i(t) \in \bigcup_{k=1}^{m-1} I_k$, $i = \overline{1, n-1}$. If $0 < \alpha_i^{(1)}(t) < \frac{1}{1+\varepsilon}$ for $t \in [0, T)$, $i = \overline{1, n-1}$, then the mentioned above inclusion is satisfied on $[0, T)$. Such inclusion enable the application of the step method, known from the theory of functional-differential equations, for construction of the solution $x(t)$. Here readers may refer to the book of [Smith (2010)].

For construction of the element $x_0(t) \in \mathcal{C}_{([0,h],E_1)}$ we construct the sequence $\{x_0^n(t)\}$:

$$x_0^n(t) = -Ax_0^{n-1} - Kx_0^{n-1} + \overline{f}(t),$$
$$x_0^0(t) = \overline{f}(t), \ t \in [0, h].$$

Because of our choice of h we have the following bound $\|A + K\|_{\mathcal{L}(\mathcal{C}_{([0,h];E_1)} \to \mathcal{C}_{([0,h];E_1)})} < 1$.

Therefore for $t \in [0, h]$ there exists the unique solution $x_0(t)$ of the equation (4.2.3). The sequence $x_0^n(t)$ converge uniformly to this solution. Let us continue the process of the desired solution construction for $t \geq h$, i.e., on I_n, $n = 1, 2, \ldots$. For sake of clarity let $\varepsilon = 1$ in (4.2.4).

Then once we have constructed the element $x_0(t) \in \mathcal{C}_{([0,h];E_1)}$ we search for the element $x_1(t)$ in the space $\mathcal{C}_{([h,2h];E_1)}$ of continuous vector functions. Let us find $x_1(t)$ from the following VIE of the second kind

$$x(t) + \int_h^t K_n^{-1}(t,t) K_t'(t,s) x(s) \, ds$$

$$= \overline{f}(t) - Ax_0 - \int_0^h K_n^{-1}(t,t) K_t'(t,s) x_0(s) \, ds$$

using the successive approximations. Here $x_0(h) = x_1(h)$.

We introduce the continuous function

$$\overline{x}_1(t) = \begin{cases} x_0(t), & 0 \leq t \leq h, \\ x_1(t), & h \leq t \leq 2h, \end{cases} \tag{4.2.5}$$

which is the contraction of the desired solution $x(t)$ onto $[0, 2h]$. Then the element $x_2(t) \in \mathcal{C}_{([2h,3h];E_1)}$ can be computed with successive approximations from the VIE of the second kind

$$x(t) + \int_{2h}^t K_n^{-1}(t,t) K_t'(t,s) x(s) \, ds$$

$$= \overline{f}(t) - A\overline{x}_1 - \int_0^{2h} K_n^{-1}(t,t) K_t'(t,s) \overline{x}_1(s) \, ds.$$

Therefore on the second step we obtained the desired solution for $t \in [0, 2h]$. We continue this process. In order to construct the desired solution $x(t) \in \mathcal{C}_{([0,T];E_1)}$ to the equation (4.1.3) on the whole interval $[0, T]$ it is enough to make N steps, where $N \geq [\frac{T}{h}] + 1$. $\qquad\square$

Remark 4.1. The problem of solution of equation (4.1.3) is well-posed on the pair of spaces $(\overset{o(1)}{\mathcal{C}}_{([0,T],E_1)}, \mathcal{C}_{([0,T],E_2)})$ if conditions of the theorem 4.1 are satisfied. This follows from Bounded Inverse Theorem (here readers may refer to chapter 12 in the monograph of Trenoguine, V. A. (1985). Analyse fonctionnelle).

Example 4.1. Integral equation

$$\int_0^{t/2} K_1(t-s)x(s)\,ds + \int_{t/2}^t K_2(t-s)x(s)\,ds = f(t),\ 0 \leq t \leq T,$$

$f(0) = 0$, $K_1(t-s) = K_2(t-s) + E$, K_1, K_2 are matrices $m \times m$, E is the unit matrix,

$$|K_2^{-1}(0)|_{\mathcal{L}(\mathbb{R}^m \to \mathbb{R}^m)} < 2,$$

matrix $K_2(t)$ and vector-function $f(t) = (f_1(t), \ldots, f_m(t))'$ have continuous derivatives w.r.t. t, and satisfy conditions of the Theorem 4.1 and has a unique continuous solution.

Example 4.2. Boundary problem

$$\begin{cases} \int_0^{t/2} \left(\dfrac{\partial^2 x(t,y)}{\partial y^2} + x(t,y) \right) dt + \int_{t/2}^t \dfrac{\partial^2 x(t,y)}{\partial y^2}\,dt = f(t,y),\ 0 \leq t \leq T,\ 0 \leq y \leq 1, \\ x(t,0) = 0, \quad x(t,1) = 0, \end{cases}$$

where the function $f(t, y)$ is continuous w.r.t. y and has continuous derivative w.r.t. t, $f(0, y) = 0$, satisfy the conditions of the Theorem 4.1. The desired continuous solution can be constructed as solution to the equivalent equation

$$x(t,y) = -\frac{1}{2} \int_0^t G(y,\xi)x(t,\xi)\,d\xi + \int_0^1 G(y,\xi)f'_t(t,\xi)\,d\xi$$

with a contractive operator where

$$G(y,\xi) = \begin{cases} (\xi-1)y, & y \leq \xi, \\ (y-1)\xi, & \xi \leq y \end{cases}$$

based on the successive approximations method.

4.3 Construction of the Asymptotic Approximation of Parametric Family of Solutions

Let the following condition be satisfied

B. There exists operator polynomials $\mathcal{P}_i = \sum\limits_{\nu+\mu=1}^{N} K_{i\nu\mu} t^\nu s^\mu$, $i = \overline{1, n}$, where $K_{i,\nu,\mu} \in \mathcal{L}(E_1 \rightarrow E_2)$, are linear continuous operators, vector-function $f^N(t) = \sum\limits_{\nu=1}^{N} f_\nu t^\nu$, polynomials $\alpha_i^N(t) = \sum\limits_{\nu=1}^{N} \alpha_{i\nu} t^\nu$, $i = \overline{1, n-1}$, where $0 < \alpha_{11} < \alpha_{21} < \alpha_{21} < \cdots < \alpha_{n-1,1} < 1$, are such as for $t \rightarrow +0$, $s \rightarrow +0$ the following estimates are satisfied

$$\|K_i(t, s) - \mathcal{P}_i(t, s)\|_{\mathcal{L}(E_1 \rightarrow E_2)} = \mathcal{O}((t+s)^{N+1}), \ i = \overline{1, n},$$

$$\|f(t) - f^N(t)\|_{E_2} = \mathcal{O}(t^{N+1}),$$

$$|\alpha_i(t) - \alpha_i^N(t)| = \mathcal{O}(t^{N+1}), \ i = \overline{1, n-1}.$$

Expansion of powers t, s in the condition **B** we call as Taylor polynomials of the corresponding elements.

Let us introduce j-parametric family of linear operators

$$B(j) = K_n(0, 0) + \sum_{i=1}^{n-1} (\alpha_i'(0))^{1+j} (K_i(0, 0) - K_{i+1}(0, 0)),$$

$j \in [0, \infty)$. Operator $B(j)$ which corresponds to the main "functional" part of the equation (4.2.1) We denote the operator $B(j)$ as *characteristic operator* of equation 4.2.1.

Let us follow the first paragraph of this section and consider equation (4.2.1) which is equivalent to the equation (4.1.3). In contrast to the first paragraph here we do not suppose that $D(0) < 1$ (see condition **A.**). Therefore the solution to the integral-functional equation (4.2.1) can be non-unique. We refer to the article of [Sidorov (2011b)] and seek an asymptotic approximation of the particular solution of the non-homogeneous equation (4.2.1) as the following polynomial

$$\hat{x}(t) = \sum_{j=0}^{N} x_j (\ln t) t^j. \tag{4.3.1}$$

Let us demonstrate that the coefficients x_j with ranges in E_1 in the general non-regular case depends on $\ln t$ and the free parameters. This is in line with the existence of nontrivial solution of the homogeneous equation.

There are regular and non-regular cases when it comes to the determination of the coefficients x_j.

Definition 4.1. j^* is the *regular point* of operator $B(j)$, if $B(j^*)$ has a bounded inverse and is a *non-regular point* otherwise.

Regular Case: Characteristic Operator $B(j)$ has Bounded Inverse Operator for Positive Integers

In this case the coefficients x_j are constant vectors from E_1. Indeed, let us substitute (4.3.1) into the equation (4.2.1). Then using the method of successive approximations and condition **B** we get the recurrent sequence of linear equations w.r.t. x_j :

$$B(0)x_0 = f'(0), \qquad (4.3.2)$$

$$B(j)x_j = M_j(x_0, \dots, x_{j-1}), \; j = 1, \dots, N. \qquad (4.3.3)$$

The vector M_j can be obtained via solutions $x_0, \dots, x_{j-1}$ and coefficients of the Taylor polynomials from the condition **B**.

Since the operators $B(j)$ are invertible in the regular case then the vectors $x_0, \dots, x_N$ can be uniquely defined and the asymptotic (4.3.1) can be constructed.

Non-regular Case: Operator $B(j)$ for Positive Integers has Non-regular Points

Let us introduce some definitions:

Definition 4.2. *Value j^* is the simple singular Fredholm point of the operator $B(j)$, if $B(j^*)$ is the Fredholm operator (readers may refer here to the textbook of [Trénoguine (1985)] and to the textbook of [Kreyszig (1978)] for more details), $\det\left[\langle B^{(1)}(j^*)\phi_i, \psi_k\rangle\right]_{i,k=1}^r \neq 0$, where $\{\phi_i\}_1^r$ is basis in $N(B(j^*))$, $\{\psi_i\}_1^r$ is a basis in $N(B'(j^*))$, $B'(j^*)$ is conjugate operator, $B^{(1)}(j)$ is derivative w.r.t. j, computed with $j = j^*$.*

Definition 4.3. *Let $B(j^*)$ be Fredholm operator, j^* we call the singular Fredholm point of index $k+1$ if $N(B(j^*)) \subset \bigcap_{i=1}^{k} N(B^{(i)}(j^*))$,*

$$\det\left[< B^{(k+1)}(j^*)\phi_i, \psi_k >\right]_{i,k=1}^r \neq 0, \; k \geq 1.$$

It is to be noted that $B^{(k)}(j) = \sum_{i=1}^{n-1} (\alpha_i'(0))^{1+j} a_i^k (K_i(0,0) - K_{i+1}(0,0))$,
where $a_i = \ln \alpha_i'(0)$.

Remark 4.1. According to the definition 4.3 the index of the simple Fredholm point is 1. If $E_1 = E_2 = \mathbb{R}^1$ then $B(j)$ is a function of j. In this case the definition 4.2 means that j^* is a single root of the equation $B(j) = 0$. The definition 4.3 means that j^* is $(k+1)$ multiple root of this equation.

Let us demonstrate that in non-regular case the coefficients x_j will be a polynomial of power $\ln t$ and depends on arbitrary constants. The order of polynomials and number of arbitrary constants are connected with indexes of singular points of the operators $B(j)$ and dimension $N(B(j))$.

Indeed since the coefficient x_0 in the non-regular case can depends on $\ln t$, then based on the method of undetermined coefficients x_0 can be determined as the solution to the difference equation

$$K_n(0,0)x_0(z) + \sum_{i=1}^{n-1} \alpha_i'(0)(K_i(0,0) - K_{i+1}(0,0))x_0(z+a_i) = f'(0), \quad (4.3.4)$$

where $a_i = \ln \alpha'(0)$, $z = \ln t$.

Here there are three cases:

Case I.

The operator $B(0)$ has a bounded inverse operator. Then the coefficient x_0 does not depends on z and can be determined uniquely from (4.3.2).

Case II.

Let $j = 0$ be the simple Fredholm point of the operator $B(j)$. Let us search the coefficient $x_0(z)$ from the difference equation (4.3.4) as a linear vector-function

$$x_0(z) = x_{01}z + x_{02}. \quad (4.3.5)$$

We substitute (4.3.5) into (4.3.4) and we get the following two equations for determination of vectors x_{01}, x_{02}

$$B(0)x_{01} = 0, \quad (4.3.6)$$

$$B(0)x_{02} + B^{(1)}(0)x_{01} = f'(0). \quad (4.3.7)$$

Let $\{\phi_i\}_1^r$ be basis in $N(B(0))$. Then $x_{01} = \sum_{k=1}^r c_k \phi_k$. Based on Fredholm alternative, the vector $c = (c_1, \ldots, c_r)'$ can be determined uniquely from the system of linear algebraic equations

$$\sum_{k=1}^r \langle B^{(1)}(0)\phi_k, \psi_i \rangle c_k = \langle f'(0), \psi_i \rangle, \quad i = \overline{1, r}$$

with nonsingular matrix. Here $\{\psi_i\}_1^r$ is a basis in $N(B'(j^*))$, $B'(j^*)$ is conjugate operator. Further the coefficient x_{02} can be determined from the equation (4.3.7) with accuracy up to $\mathrm{span}(\phi_1, \ldots, \phi_r)$. For this reason we employ the formula

$$x_{02} = \sum_{k=1}^{r} d_k \phi_k + \Gamma(f'(0) - B^{(1)} x_{01}),$$

where

$$\Gamma = (B(0) + \sum_{k=1}^{r} \langle \cdot, \gamma_k \rangle z_k)^{-1}$$

is Trenogin-Schmidt operator (readers may refer to the textbooks of [Trénoguine (1985)], [Zeidler (1986)]), $d_1, \ldots, d_r$ are arbitrary constants. Therefore in the 2nd case the coefficient $x_0(z)$ is linear w.r.t. z and depends on r arbitrary constants.

Case III. Let $j = 0$ be the singular Fredholm point of the operator $B(j)$ of index $k + 1$, $k \geq 1$. The solution $x_0(z)$ to the difference equation (4.3.4) we seek as the following polynomial

$$x_0(z) = x_{01} z^{k+1} + x_{02} z^k + \cdots + x_{0k+1} z + x_{0k+2}. \qquad (4.3.8)$$

Let us substitute the polynomial (4.3.8) into the system (4.3.4) and let us take into account the following identity

$$\frac{d^k}{dj^k} B(j) = \sum_{i=1}^{n-1} (\alpha_i'(0))^{1+j} a_i^k (K_i(0,0) - K_{i+1}(0,0)),$$

where $a_i = \ln \alpha_i'(0)$. Let us match the coefficients on powers of $z^{k+1}, z^k, \ldots, z, z^0$ to zero. As a result we get the sequence of linear operator equations w.r.t. $x_{01}, x_{02}, \ldots, x_{0k+2}$:

$B(0) x_{01} = 0,$

$B(0) x_{02} + B^{(1)}(0) \begin{pmatrix} k+1 \\ k \end{pmatrix} x_{01} = 0,$

$B(0) x_{0l+1} + B^{(l)}(0) \begin{pmatrix} k+1 \\ k+1-l \end{pmatrix} x_{01} + B^{(l-1)}(0) \begin{pmatrix} k \\ k+1-l \end{pmatrix} x_{02}$

$\quad + \cdots + B^{(1)}(0) \begin{pmatrix} k+1-l+1 \\ k+1-l \end{pmatrix} x_{0l} = 0, \ l = 1, \ldots, k,$

$B(0) x_{0k+2} + B^{(k+1)}(0) x_{01} + B^{(k)}(0) x_{02} + \ldots B^{(1)}(0) x_{0k+1} = f'(0). \quad (4.3.9)$

Let us follow the definition 4.3 here. Therefore we have the following inclusion

$$N(B(0)) \subset \bigcap_{i=1}^{k} N\left(\frac{d^i B(j)}{dj^i} \bigg|_{j=0} \right).$$

Hence $B^{(i)}(0)x_{0i+1} = 0$, $i = \overline{0,k}$ and coefficients $x_{01}, \ldots, x_{0k+1}$ can be determined from the homogeneous equation $B(0)x = 0$ based on the formulas $x_{0i} = \sum_{j=1}^{r} c_{ij}\phi_j$, $i = \overline{1, k+1}$. Therefore the equation (4.3.9) becomes

$$B(0)x_{0,k+2} + B^{(k+1)}(0)x_{01} = f'(0). \tag{4.3.10}$$

Since $B(0)$ is the Fredholm operator and

$$\det\Big[< B^{(k+1)}(0)\phi_i, \psi_k > \Big]_{i,k=\overline{1,r}} \neq 0,$$

then the vector $c^1 \overset{\text{def}}{=} (c_{11}, \ldots, c_{1r})'$ can be determined uniquely from the conditions of resolvability of the equation (4.3.10). Therefore we have

$$x_{0\,k+2} = \sum_{j=1}^{r} c_{k+2\,j}\phi_j + \hat{x}_{k+2},$$

$\hat{x}_{k+2}$ is particular solution to the equation (4.3.10). Vector $c^{k+2} \overset{\text{def}}{=} (c_{k+2,1}, \ldots, c_{k+2,r})'$, as well as $c^i = (c_{i1}, \ldots, c_{ir})', i = \overline{2, k+1}$, remains arbitrary. As result, in Case III the coefficient $x_0(z)$ is polynomial of $k+1$th power w.r.t. z and depends on $r \times (k+1)$ arbitrary constants.

Let us apply the method of undetermined coefficients and take into account the equality

$$\int t^j \ln^k t\, dt = t^{j+1} \sum_{s=0}^{k} (-1)^s \frac{k(k-1)\ldots(k-(s-1))}{(j+1)^{s+1}} \ln^{k-s} t.$$

We can construct the difference equations for determination of the coefficient $x_1(z)$ ($z = \ln t$) and the next coefficients of the asymptotic approximation (4.3.1). Indeed, let us take into account the definition of the operator F (see the equation (4.2.1)). Then we get

$$F(x)\Big|_{x=x_0(z)+x_1(z)t} = \Big[K_n(0,0)x_1(z) + \sum_{i=1}^{n-1} (\alpha_i'(0))^2 (K_i(0,0)$$

$$- K_{i+1}(0,0))x_1(z+a_i) + P_1(x_0(z))\Big]t + r(t),$$

$$\tag{4.3.11}$$

with estimate $r(t) = o(t)$. Here $P_1(x_0(z))$ is a certain polynomial w.r.t. z. Its power is equal to index of the singular Fredholm point $j = 0$ of the operator $B(j)$. Taking into account (4.3.11) and the estimate $r(t) = o(t)$

as $t \to 0$ we conclude that the coefficient $x_1(z)$ has to satisfy the following difference equation

$$K_n(0,0)x_1(z) + \sum_{i=1}^{n-1}(\alpha'(0))^2 \big(K_i(0,0) - K_{i+1}(0,0)\big)x_1(z + a_i)$$

$$+ P_1(x_0(z)) = 0. \tag{4.3.12}$$

If $j = 1$ is the regular point of the operator $B(j)$ then the equation (4.3.12) has solution $x_1(z)$ as polynomial of the same order as index of the singular Fredholm point $j = 0$ of the operator $B(0)$. If $j = 1$ is the singular Fredholm point of the operator $B(j)$ the solution $x_1(z)$ can be constructed as the polynomial of power $k_0 + k_1$, where k_0 and k_1 are indexes of the corresponding singular Fredholm points $j = 0$, $j = 1$ of the operator $B(j)$. Coefficient $x_1(z)$ will depends on $r_0 k_0 + r_1 k_1$ arbitrary constants, where $r_0 = \dim N(B(0))$, $r_1 = \dim N(B(1))$.

Let us introduce the condition

C. Let operator $B(j)$ for $j \in (0, 1, \ldots, N)$ has regular points only or singular Fredholm points $j_1, \ldots, j_\nu$ of indexes k_i, $\dim N(Bj_i) = r_i$, $i = \overline{1, \nu}$.

Then by the similar means we can determine the rest of the coefficients $x_2(z), \ldots, x_N(z)$ in $\hat{x}(t)$ from the sequence of difference equations

$$K_n(0,0)x_j(z) + \sum_{i=1}^{n-1}(\alpha'(0))^{1+j}\big(K_i(0,0) - K_{i+1}(0,0)\big)x_j(z + a_i)$$

$$+ P_j(x_0(z), \ldots, x_{j-1}(z))) = 0, \; j = \overline{2, N}.$$

Therefore we have the following

Lemma 4.1. *Let the conditions* **B** *and* **C** *be satisfied. Then exists the vector-function* $\hat{x}(t) = \sum\limits_{i=0}^{N} x_i(\ln t)t^i$, *such as* $\|F(\hat{x}(t))\|_{E_2} = o(t^N)$, *where operator* F *is defined with formula (4.2.1). The coefficients* $x_i(\ln t)$ *are polynomials of* $\ln t$ *of increasing powers and they are smaller than* $\sum\limits_j k_j$ *singular Fredholm points* $j \in \{0, 1, 2, \ldots, N\}$ *of characteristic operator* $B(j)$. *Coefficients* $x_i(\ln t)$ *depends on* $\sum_{j=0}^{i} \dim N(B(j))k_j$ *arbitrary constants.*

Remark 4.3. If $B(0)$ is the Fredholm operator and $\dim N(B(0)) \geq 1$, then coefficient $x_0(\ln t)$ can be linear function of $\ln t$ and vector-function $\hat{x}(t)$ will increase unboundedly as $t \to +0$ (briefly, $\hat{x} \in C_{((0,T];E_1)}$).

4.4   Theorem of Existence of Continuous Parametric Solutions

Since $0 \leq \alpha_i'(0) < 1$, $\alpha_i(0) = 0$, $i = \overline{1, n-1}$, then for any $0 < \varepsilon < 1$ exists $T' \in (0, T]$ such as $\displaystyle\max_{i=\overline{1,n-1}, t \in [0,T']} |\alpha_i'(t)| \leq \varepsilon$ and $\displaystyle\sup_{i=\overline{1,n-1}, t \in (0,T']} \frac{\alpha_i(t)}{t} \leq \varepsilon$.

Let us introduce the condition

D. Let operator $K_n(t, t)$ has inverse bounded operator for $t \in [0, T']$ and N^* is selected to satisfy the following inequality

$$\sup_{t \in (0,T')} \varepsilon^{N^*} \sum_{i=1}^{n-1} |\alpha_i^{(1)}(t)| \left\| K_n^{-1}(t, t) \Big(K_i(t, \alpha_i(t)) \right.$$

$$\left. - K_{i+1}(t, \alpha_i(t)) \Big) \right\|_{\mathcal{L}(E_1 \to E_1)} \leq q < 1.$$

Lemma 4.2. *Let condition* **D** *be fulfilled. Let in* $C_{([0,T'];E_1)}$ *exists element* $\hat{x}(t)$ *such as for* $t \to +0$

$$\|F(\hat{x}(t))\|_{E_2} = o(t^N), \, N \geq N^*.$$

Then equation (4.1.3) in $C_{([0,T'];E_1)}$ *possesses a solution as follows*

$$x(t) = \hat{x}(t) + t^{N^*} u(t), \tag{4.4.1}$$

where $u(t)$ *is unique and defined with successive approximations.*

Proof. Let us substitute (4.4.1) into the equation (4.2.1). We get the following integral-functional equation for determination of function $u(t)$

$$K_n(t, t) u(t) + \sum_{i=1}^{n-1} \alpha_i'(t) \left(\frac{\alpha_i(t)}{t} \right)^{N^*} \Big(K_i(t, \alpha_i(t))$$

$$- K_{i+1}(t, \alpha_i(t)) \Big) u(\alpha_i(t))$$

$$+ \sum_{i=1}^{n} \int_{\alpha_{i-1}(t)}^{\alpha_i(t)} K_i^{(1)}(t, s) \left(\frac{s}{t} \right)^{N^*} u(s) \, ds + F(\hat{x}(t))/t^{N^*} = 0. \tag{4.4.2}$$

Let us introduce the linear operators

$$Lu \overset{\text{def}}{=} K_n^{-1}(t,t) \sum_{i=1}^{n-1} \alpha_i'(t) \left(\frac{\alpha_i(t)}{t}\right)^{N^*} \Big\{ K_i(t, \alpha_i(t))$$

$$- K_{i+1}(t, \alpha_i(t)) \Big\} u(\alpha_i(t)),$$

$$Ku \overset{\text{def}}{=} \sum_{i=1}^{n} \int_{\alpha_{i-1}(t)}^{\alpha_i(t)} K_n^{-1}(t,t) K_i^{(1)}(t,s)(s/t)^{N^*} u(s)\, ds.$$

Then system (4.4.2) can be rewritten as follows

$$u + (L + K)u = \gamma(t),$$

where $\gamma(t) = K_n^{-1}(t,t) F(\hat{x}(t))/t^{N^*}$ is continuous vector-function. Let us introduce a Banach space X of continuous functions $u(t)$ with ranges in a Banach space E_1 and the following norm

$$||u||_l = \max_{0 \le t \le T'} e^{-lt} ||u(t)||_{E_1}, \, l > 0.$$

Norm of linear functional operator L satisfies the following bound

$$||L||_{\mathcal{L}(X \to X)} \le q$$

because of inequality $\displaystyle\sup_{t \in (0, T']} \frac{\alpha_i(t)}{t} \le \varepsilon < 1$ and based on the condition **D** $\forall l \ge 0$. Moreover for integral operator K the following bound is satisfied

$$||K||_{\mathcal{L}(X \to X)} \le q_1 < 1 - q$$

for big enough l. As results, for big enough $l > 0$ we have

$$||L + K||_{\mathcal{L}(X \to X)} < 1,$$

i.e., the linear operator $L + K$ is contractive operator in space X. Therefore the sequence $\{u_n\}$ converges. Here $u_n = -(L + K)u_{n-1} + \gamma(t)$, $u_0 = \gamma(t)$. $\square$

Theorem 4.2 (The Main Theorem). *Let conditions* **B**, **C**, **D**, *and* $f(0) = 0$ *be satisfied. Let operator* $B(0)$ *has bounded inverse operator. Then equation (4.1.3) in space* $\mathcal{C}_{([0,T];E_1)}$, $0 \le t \le T' \le T$ *possesses solution*

$$x(t) = \hat{x}(t) + t^{N^*} u(t),$$

which depends on $\displaystyle\sum_{i=1}^{\nu} r_i k_i$ *arbitrary constants. Moreover, element* $\hat{x}$ *can be constructed as logarithmic-power sum (4.3.1), and* $u(t)$ *is uniquely computed with successive approximations. The asymptotic estimate* $||x(t) - \hat{x}(t)||_{E_1} = \mathcal{O}(t^{N^*})$ *is satisfied as* $t \to +0$.

Proof. Let us employ Lemma 4.1 and take into account the conditions of the theorem. Therefore the construction of asymptotic approximation $\hat{x}(t)$ of the desired solution as following logarithmic-power polynomial

$$\sum_{i=0}^{N} x_i(\ln t)t^i$$

is possible. And coefficients $x_i(\ln t)$ depend on certain number of arbitrary constants. We can now apply Lemma 4.2 and $x(t) = \hat{x}(t) + t^{N^*}u(t)$. Therefore continuous function $u(t)$ can be constructed with successive approximations. Theorem is proved. □

The parametric family of solutions constructed on $[0, T']$ we can continued with "step method" on the interval $[0, T]$.

If $j = 0$ is Fredholm point of operator $B(j)$ and $\dim N(B(0)) \geq 1$ then based on Remark 4.2 the coefficient $x_0(\ln t)$ in the asymptotic $\hat{x}(t)$ can be linear function of $\ln t$. In this case the solution $x(t) \in C_{((0,T];E_1)}$ and grow unbound as $t \to +0$.

Example 4.3. Equation

$$\int_0^{t/2} \int_0^1 K(y, y_1)x(s, y_1)\,dy_1 ds$$

$$+ \int_{t/2}^t \left(\int_0^1 K(y, y_1)x(s, y_1)\,dy_1 - 2x(s, y) \right) ds = g(y)t,$$

where $0 < t < \infty$, $0 < y < 1$, 1 is eigenvalue of continuous symmetric kernel $K(y, y_1)$ of rank r, $\{\phi_1(y), \ldots, \phi_r(y)\}$ is corresponding orthonormal system of eigenfunctions on $[0, 1]$, $g(y) \in C_{[0,1]}$, meets the conditions of Theorem 4.2. In such case $j = 0$ is the simple singular Fredholm point of corresponding characteristic operator $B(j)$. The solution of equation is following

$$x(t, y) = -\frac{\ln t}{\ln 2} \sum_{i=1}^{r} \int_0^1 \phi_i(y)\phi_i(y_1)f(y_1)\,dy_1 + c_1\phi_1(y) + \ldots + c_r\phi_r(y) + x_0(y),$$

$c_1, \ldots, c_r$ are arbitrary constants, $x_0(y)$ is the particular solution of the Fredholm integral equation of the second kind

$$x(y) = \int_0^1 K(y, y_1)x(y_1)\,dy_1 - f(y) + \sum_{i=1}^{r} \phi_i(y) \int_0^1 \phi_i(y_1)f(y_1)\,dy_1.$$

Example 4.4. Let L_n and L_m be the following linear differential operators

$$L_n[u] \equiv \frac{\partial^n u(x,s)}{\partial x^n} + a_{n-1}(x)\frac{\partial^{n-1} u(x,s)}{\partial x^{n-1}} + \cdots + a_0(x)u(x,s),$$

$$L_m[u] \equiv b_m(x,t,s)\frac{\partial^m u(x,s)}{\partial x^m} + b_{m-1}(x,t,s)\frac{\partial^{m-1} u(x,s)}{\partial x^{m-1}}$$

$$+ \cdots + b_0(x,t,s)u(x,s), \ m \leq n-2.$$

Here a_i, b_i are known continuous coefficients, b_i have continuous derivatives
w.r.t. t for $a \leq x, s \leq b$, $0 \leq t \leq T$. Consider the boundary conditions

$$V_k[u] \equiv \sum_{i=0}^{n-1}\left(\alpha_{ki}\frac{\partial^i u(x,s)}{\partial x^i}\bigg|_{x=a} + \beta_{ki}\frac{\partial^i u(x,s)}{\partial x^i}\bigg|_{x=b}\right) = 0, \ k = \overline{1,n}.$$

We assume that the homogeneous boundary value problem

$$L_m[u] = 0, \ V_k[u] = 0, \ k = \overline{1,n}$$

has only the trivial solution and let us introduce the corresponding integral
Green operator

$$G = \int_a^b G(x,s)[\cdot]\,ds.$$

Let us address the Volterra integral equation

$$\int_0^{\alpha t} L_m[u]\,ds + \int_{\alpha t}^t L_n[u]\,ds = f(t,x)$$

with condition

$$V_k[u] = 0, \ k = \overline{1,n},$$

where $\alpha \in (0,1)$, $f(t,x)$ is continuous function differentiable w.r.t. t for
$x \in [a,b]$, $0 \leq t \leq T$, $f(0,x) = 0$. Let

$$\alpha\left(1 + \max_{0 \leq x \leq b}\int_a^b |L_m[G(x,s)]|\,ds\right)\bigg|_{t=0} \leq q < 1.$$

Then after the substitution $u(x,t) = \int_a^b G(x,s)\omega(s,t)\,ds$ we are getting
th equation w.r.t. ω which meets the conditions of Theorem 4.2. Based
on Theorem 4.2 considered boundary value problem has a unique classical
solution if

$$0 < \alpha < \left(1 + \max_{0 \leq x \leq b}\int_a^b |L_m[G(x,s)]|\,ds\bigg|_{t=0}\right)^{-1}.$$

4.5 Final Remarks on Generalized Solutions

In case of single equation for $E_1 = E_2 = \mathbb{R}^1$, our method for solution of the difference systems turns to be known the A. O. Gelfond's method of construction of particular solutions of inhomogeneous difference equations with polynomial right-hand side (here readers may refer to the book of [Gelfond (1971)], p.338). We employed the results from functional analysis presented in the article of [Sidorov and Sidorov (2006)] and the ideas of the Gelfond's method in the theory of the Volterra linear equations of the first kind with piecewise continuous operators acting on a Banach spaces. If $f(0) \neq 0$ then the equation (4.1.3) does not possess solution in class of continuous functions. In this case our method and results from the paper of [Sidorov (2011b)] enable construction of the solutions in class of distributions. Here readers may also refer to the textbooks of [Kanwal (2013)], [Vladimirov (1979)]. The construction of the solutions in class of distributions in sense of Sobolev-Schwartz theory is considered in the next chapter.

Chapter 5

Generalized Solution to the Volterra Equations with Piecewise Continuous Kernels and Sources

5.1 Problem Statement. The State of the Art

Let us define the triangular region $D = \{s, t; 0 < s < t < T\}$ and introduce the functions $s = \alpha_i(t), i = \overline{1, n}$, which are continuous and have continuous derivatives for $t \in (0, T)$. We suppose $\alpha_i(0) = 0$, $0 < \alpha_1(t) < \cdots < \alpha_{n-1}(t) < t$ for $t \in (0, T)$, $0 < \alpha_1'(0) < \cdots < \alpha_{n-1}'(0) < 1$, and functions $s = \alpha_i(t)$, $i = \overline{0, n}$, $\alpha_0(t) = 0$, $\alpha_n(t) = t$, split the region D into the following disjoint sectors $D_1 = \{s, t : 0 \leq s < \alpha_1(t)\}$, $D_i = \{s, t : \alpha_{i-1}(t) < s < \alpha_i(t), i = \overline{2, n}\}$, $\overline{D} = \bigcup_{1}^{n} D_i$. Let us introduce the continuous functions $K_i(t, s)$ defined for $t, s \in D_i$, and differentiable w.r.t. t, $i = \overline{1, n}$.

Let us consider the integral operator

$$\int_0^t K(t, s)u(s)ds \stackrel{\text{def}}{=} \sum_{i=1}^{n} \int_{\alpha_{i-1}(t)}^{\alpha_i(t)} K_i(t, s)u(s)ds \tag{5.1.1}$$

with piecewise continuous kernels

$$K(t, s) = \begin{cases} K_1(t, s), & t, s \in D_1, \\ \cdots & \cdots \\ K_n(t, s), & t, s \in D_n. \end{cases} \tag{5.1.2}$$

In this chapter we continue our studies of VIE

$$\int_0^t K(t, s)u(s)\, ds = f(t),\ 0 < t < T \leq \infty, \tag{5.1.3}$$

where function $f(t)$ has a continuous derivative for $t \in (0, T)$, $f(0) \neq 0$. In addition we consider equation (5.1.1) with piecewise continuous function $f(t)$.

VIE (5.1.3) does not possesses classic solutions if $f(0) \neq 0$. In this chapter we construct the generalized solution of VIE (5.1.3) in the space of Sobolev-Schwartz distributions.

As we outlined in the previous chapters, the differentiation of VIE (5.1.3) leads to integral-functional equation and its solution is not unique in the general case. That is why study of VIE (5.1.3) cannot be performed using only the classic methods in the Volterra theory. We consider the equation (5.1.3) using the elementary results of the theory of integral and difference equations, functional analysis, Sobolev-Schwartz distributions and theory of functional equations with perturbed argument of neutral type introduced in the papers of [Sidorov and Trufanov (2009); Sidorov *et al.* (2007)]. For systematic treatment of functional equations with causal operators readers may refer to classic book of [Corduneanu (2002)].

We confine ourselves to outlining the formalism of the generalized solutions construction for equation (5.1.1) in the form $u(t) = a\delta^+(t) + x(t)$, where $x(t) \in C_{(0,T)}$, $\delta^+(t)$ is so-called the *right* delta-function.

Let us also introduce the right Heaviside function as follows

$$\theta^+(s) := \begin{cases} 1, \ s > 0, \\ 0, \ s \leq 0, \end{cases} \tag{5.1.4}$$

and corresponding the right delta-function δ^+ with following properties

(1) $M(s)\delta^+(s) = M(+0)\delta^+(s)$ for $\forall M(s) \in C_{(0,T]}$, where $M(+0) = \lim\limits_{s \to +0} M(s)$.

(2) $\int_0^t \delta^+(s)\,ds = \theta^+(t)$.

Let $0 = \alpha_0(t) < \alpha_1(t) < \cdots < \alpha_n(t) = t$, and let $\alpha_i(t)$ be continuous functions, $\alpha_i(0) = 0$. Then for the 2nd property it follows that

$$J_i(t) \equiv \int_{\alpha_{i-1}(t)}^{\alpha_i(t)} \delta^+(s)\,ds$$

$$= -\int_0^{\alpha_{i-1}(t)} \delta^+(s)\,ds + \int_0^{\alpha_i(t)} \delta^+(s)\,ds = \theta^+(\alpha_i(t)) - \theta^+(\alpha_{i-1}(t)).$$

Then

$J_i(0) = 0, i = 1, \cdots, n$, $J_1(t) = 1$ for $t > 0$, $J_i(t) = 0$ for $i = \overline{2, n}$ and $t > 0$.

Such left and right delta-functions were also used in the paper of [Toshio *et al.* (1977)].

So, let's look for the solution of the equation (5.1.1) in the form

$$u(s) = a\delta^+(s) + x(s), \text{ where } x(s) \in C_{(0,T)}. \tag{5.1.5}$$

Then formal substitution of (5.1.5) into equation (5.1.1) leads us to the equation

$$aK_1(t, +0) + \int_0^t K(t, s)x(s)\, ds = f(t), \ t > 0.$$

Let a be selected as $\frac{f(+0)}{K_1(+0,+0)}$ and we get the following integral equation w.r.t. the regular component $x(t)$ of the generalized solution (5.1.5)

$$\int_0^t K(t, s)x(s)\, ds = \hat{f}(t), \ t > 0. \tag{5.1.6}$$

Here

$$\hat{f}(t) = f(t) - K_1(t, +0)\frac{f(+0)}{K_1(+0, +0)}.$$

Because $\hat{f}(+0) = 0$ then the calculation of the right derivative of both parts of (5.1.6) leads to the equivalent equation of the second kind

$$K_n(t, t)x(t) + \sum_{i=1}^{n-1} \alpha_i'(t)\{K_i(t, \alpha_i(t)) - K_{i+1}(t, \alpha_i(t))\}x(\alpha_i(t))$$

$$+ \sum_{i=1}^{n} \int_{\alpha_{i-1}(t)}^{\alpha_i(t)} \frac{\partial}{\partial t} K_i(t, s)x(s)\, ds = f'(t) - \frac{d}{dt}K_1(t, +0)\frac{f(+0)}{K_1(+0, +0)}.$$

Let us rewrite the last equation in a more compact form by introducing two linear operators

$$Ax := \sum_{i=1}^{n-1} K_n^{-1}(t, t)\alpha_i'(t)\{K_i(t, \alpha_i(t)) - K_{i+1}(t, \alpha_i(t))\}x(\alpha_i(t)),$$

$$Kx := \sum_{i=1}^{n} \int_{\alpha_{i-1}(t)}^{\alpha_i(t)} K_n^{-1}(t, t)\frac{\partial}{\partial t}K_i(t, s)x(s)\, ds.$$

As a result, in order to define a regular component of a generalized solution of equation (5.1.1) we got the following equation of the second kind

$$x(t) + Ax + Kx = K_n^{-1}(t, t)\left(f'(t) - \frac{d}{dt}K_1(t, +0) \right)\frac{f(+0)}{K_1(+0, +0)}.$$

To study this equation one can use the methods presented in chapter 1.

Example 5.1.

$$\int_0^{t/2} x(s)ds + 2 \int_{t/2}^t x(s)ds = 2 + t,\ t > 0.$$

An equivalent equation in this example is as follows $-\frac{t}{2}x(\frac{t}{2}) + 2x(t) = 2\delta(t) + 1$. The desired solution is as follows $x(t) = 2\delta(t) + 2/3$.

Example 5.2.

$$\int_0^{t/2} x(s)ds - \int_{t/2}^t x(s)ds = 1 + t,\ t > 0.$$

Here an equivalent equation is as follows $x(\frac{t}{2}) - x(t) = \delta(t) + 1$. It has c–parametric family of generalized solutions $x(t) = \delta(t) + c - \frac{\ln t}{\ln 2}$, c is constant.

*Generalizations and other approaches for generalized
solutions constructions for the Volterra integral equations of the first kind*

The proposed formalism for the generalized solutions construction does not completely disclose the structure of all the possible generalized solutions. Furthermore, it is important for applications to relax the conditions on the smoothness of functions $K_i(t, s)$, $f(t)$.

Therefore, for a more complete study it is helpful to write the problem in the form of the following equation

$$\int_0^t K(t, s)\, dg(s) = f(t),$$

where $g(s)$ is unknown function of bounded variation. In this case for calculation of the desired solution $g(s)$ one may employ the Lebesgue decomposition (here readers may refer to classic book of [Halmos (1978)])

$$g = g_c + g_d + g_s,$$

which is unique up to constants. Here g_c is an absolutely continuous function, g_d is a Jump function and g_s is the *fonction des singularités* of the continuous part of g. These functions g_s are also called singular functions and can be characterized in a few ways. For instance they are those continuous functions of bounded variation whose classical derivative vanishes almost everywhere, or those continuous functions of bounded variation whose distributional derivative is singular with respect to the Lebesgue measure.

In this case it is natural to seek a generalized solution as follows

$$g = a\mu + \nu,$$

where a is arbitrary constant, μ are ν – measures, e.g., Borel charges of bounded variation on certain intervals. This problem statement was proposed by Professor Alfredo Lorenzi.

5.2 Volterra Equations of the First Kind with Discontinuous Sources

The objective of this section is to study the scalar equation (5.1.1) with relaxed condition on the source function $f(t)$. In applications the source function can be jump discontinuous.

So, let us consider the equation

$$\mathcal{K}x = f(t), \quad t \in \mathbb{R}^1, \ t < T < \infty, \tag{5.2.1}$$

where

$$\mathcal{K}x = \sum_{i=1}^{n} \int_{\alpha_{i-1}(t)}^{\alpha_i(t)} K_i(t, s) x(s) \, ds,$$

$$f(t) = \begin{cases} f_1(t), & -\infty < t < T_1, \\ f_2(t), & T_1 < t < T_2, \\ \cdots & \cdots\cdots\cdots \\ f_m(t), & T_{m-1} < t < T < \infty, \end{cases}$$

$$0 < T_1 < T_2 < \cdots < T_{m-1}.$$

Let us again assume $K_i(t, s)$ be defined, continuous and differentiable in the domains $D_i = \{t, s | \alpha_{i-1}(t) < s < \alpha_i(t)\}$, functions $f_i(t)$ are defined, continuous and differentiable for $T_{i-1} < t < T_i$, $T_0 = 0, T_m = T$.

In the next Lemma we obtain the sufficient conditions for reduction of equation (5.2.1) to the weakly regular equation with continuous source function based on the substitution

$$x(t) = \sum_{k=1}^{m-1} c_k \delta(t - T_k) + u(t). \tag{5.2.2}$$

In (5.2.2) $\delta(t - T_k)$ is Dirac delta function, $u(t)$ is regular function, and $c_1, \ldots, c_{m-1}$ are constant values calculated using the formula (5.2.3).

Lemma 5.1. *Let functions $K_i(t, s)$ be defined in the domains D_i, piecewise continuous function $f(t)$ has exactly $m - 1$ the first-order discontinuities $T_1, \ldots, T_{m-1}$. Let $K_1(t, T_i) = K_2(t, T_i) = \cdots = K_n(t, T_i)$ for $t \in (0, T)$. Then substitution (5.2.2), where*

$$\mathbf{c} = \mathfrak{A}^{-1}\mathbf{b}, \tag{5.2.3}$$

and

$$\mathfrak{A} = \begin{pmatrix} K_1(T_1, T_1) & 0 & 0 \cdots & 0 \\ K_1(T_2, T_1) & K_1(T_2, T_2) & 0 \cdots & 0 \\ \cdots & \cdots & \cdots \cdots & \cdots \\ K_1(T_{m-1}, T_1) & K_1(T_{m-1}, T_2) & \cdots \cdots & K_1(T_{m-1}, T_{m-1}) \end{pmatrix},$$

$$\mathbf{b} = \begin{pmatrix} f_2(T_1) - f_1(T_1) \\ f_3(T_2) - f_2(T_2) \\ \cdots \\ f_m(T_{m-1}) - f_{m-1}(T_{m-1}) \end{pmatrix}$$

reduces the equation (5.2.1) to

$$\mathcal{K}u = \tilde{f}(t), \tag{5.2.4}$$

where $\tilde{f}(t) : [0, T] \to \mathbb{R}^1$ is continuous function.

Proof. The substitution (5.2.2) reduces the equation (5.2.1) to (5.2.4), where

$$\hat{f}(t) = f(t) - \sum_{i=1}^{n} \int_{\alpha_{i-1}(t)}^{\alpha_i(t)} K_i(t, s)(c_1\delta(s - T_1) + \cdots + c_{m-1}\delta(s - T_{m-1}))\,ds.$$

Let's take into account that $K_i(t, s)\delta(s - T_k) = K_i(t, T_k)\delta(s - T_k), i = \overline{1, n}$, where $K_1(t, T_k) = \ldots = K_n(t, T_k)$. Then we obtain the following equality

$$\hat{f}(t) = f(t) - \sum_{k=1}^{m} K_1(t, T_k)c_k \sum_{i=1}^{n} \int_{\alpha_{i-1}(t)}^{\alpha_i(t)} \delta(s - T_k)\,ds. \tag{5.2.5}$$

Because $\alpha_0(t) = 0 < \alpha_1(t) < \ldots < \alpha_n(t) = t$, we get the following equality

$$\sum_{i=1}^{n} \int_{\alpha_{i-1}(t)}^{\alpha_i(t)} \delta(s - T_k)\,ds = \int_0^t \delta(s - T_k)\,ds = \begin{cases} 1, & t > T_k, \\ 0, & t < T_k. \end{cases}$$

Then

$$\hat{f}(t) = f(t) - c_1 \times \left\{ \begin{array}{ll} 0, & 0 \le t < T_1 \\ K_1(t, a_1), & t > T_1 \end{array} \right\} - c_2 \times \left\{ \begin{array}{ll} 0, & t < T_2 \\ K_1(t, a_2), & t > T_2 \end{array} \right\}$$

$$- \ldots - c_m \times \left\{ \begin{array}{ll} 0, & t < T_{m-1} \\ K_1(t, a_{m-1}), & t > T_{m-1} \end{array} \right\}.$$

Vector **c** we can get from the following conditions

$$\left\{ \begin{array}{l} -K_1(T_1, T_1)c_1 + f_1(T_1) = f_2(T_1), \\ -K_1(T_2, T_1)c_1 - K_1(T_2, T_2)c_2 + f_2(T_2) = f_3(T_2), \\ \qquad \cdots\cdots\cdots\cdots\cdots \qquad \cdots\cdots\cdots\cdots \\ -K_1(T_{m-1}, T_1)c_1 - K_1(T_{m-1}, T_2)c_2 \\ -\ldots - K_1(T_{m-1}, T_{m-1})c_{m-1} + f_{m-1}(T_{m-1}) = f_m(T_{m-1}), \end{array} \right. \qquad (5.2.6)$$

which ensure continuity of the function $\hat{f}(t)$ on $[0, T]$. The condition (5.2.6) obviously coincides with the system of linear algebraic equation (5.2.3) with diagonal non-singular matrix. Then vector **c** in (5.2.2) leads (5.2.1) to equation with continuous source function. $\qquad\qquad\square$

Example 5.3. Let us consider the VIE with source function is jump discontinuous in origin

$$\int_0^t (2 + t - s)x(s)\, ds = \left\{ \begin{array}{ll} 0, & t < 0, \\ 1 + t, & 0 < t < \infty. \end{array} \right.$$

$f(0) \ne 0$. We look for the solution as $x(t) = c\delta(t) + u(t)$, where $u(t)$ is piecewise continuous bounded function which we define from the equation $\int_0^t (2 + t - s)u(s)\, ds = \hat{f}(t, c)$,

$$\hat{f}(t, c) = \left\{ \begin{array}{ll} 0, & t < 0, \\ 1 + t, & 0 < t < \infty \end{array} \right. + \left\{ \begin{array}{ll} 0, & t < 0, \\ -(2 + t)c, & 0 < t < \infty. \end{array} \right.$$

Since $\lim_{t \to -0} \hat{f}(t, c) = 0$, $\lim_{t \to +0} \hat{f}(t, c) = 1 - 2c$, then be obtain c from the equation $0 = 1 - 2c \Rightarrow c = 1/2$. Therefore the desired solution

$$x(t) = \frac{1}{2}\delta(t) + \left\{ \begin{array}{ll} 0, & -\infty < t < 0, \\ \frac{1}{4}e^{-t/2}, & 0 < t < +\infty. \end{array} \right.$$

The regular part of solution is bounded function $u(t)$ with single discontinuity in $t = 0$, $\sup\limits_{-\infty < t < \infty} u(t) = \frac{1}{4}$.

Example 5.4.

$$\int_{t/2}^{t} x(s)\,ds = \begin{cases} 0, & t < 1, \\ t - 1, & t > 1. \end{cases}$$

Here $f(t)$ is continuous differentiable function, $f'(t) = \begin{cases} 0, & t < 1, \\ 1, & t > 1. \end{cases}$ is piecewise continuous function, $f(0) = 0$. Therefore, we seek a solution in the class of piecewise continuous functions. First we construct the functional equation

$$x(t) - \frac{1}{2}x\left(\frac{t}{2}\right) = \begin{cases} 0, & t < 1, \\ 1, & t > 1. \end{cases}$$

and apply the method of successive approximations

$$x_n(t) = \frac{1}{2}x_{n-1}\left(\frac{t}{2}\right) + \begin{cases} 0, & t < 1, \\ 1, & t > 1, \end{cases}$$

where $x_0(t) = 0$, $x_1(t) = \begin{cases} 0, & t < 1, \\ 1, & t > 1, \end{cases}$ $x_2(t) = \begin{cases} 0, & t < 1, \\ 1, & 1 < t < 2, \\ 1 + 1/2, & 2 < t, \end{cases}$

$$\ldots\ldots, x_n(t) = \begin{cases} 0, & t < 1, \\ 1, & 1 < t < 2, \\ 1 + 1/2, & 2 < t < 2^2, \\ \ldots\ldots \\ \sum_{i=0}^{n-1}\left(\frac{1}{2}\right)^i, & 2^{n-1} < t. \end{cases}$$

The desired solution is following

$$x^*(t) = \begin{cases} 0, & -\infty < t < 1, \\ 1, & 1 < t < 2, \\ 1 + 1/2, & 2 < t < 2^n, \\ \ldots\ldots \\ 2 - \left(\frac{1}{2}\right)^n, & 2^{n-1} < t < 2^n, \\ \ldots\ldots \end{cases}$$

And $\sup\limits_{-\infty \le t \le \infty} x^*(t) = 2$.

Remark 5.1.

As footnote let us note that one may employ this method to study the VIE with piecewise continuous kernels $K_i(t, s)$ which are discontinuous

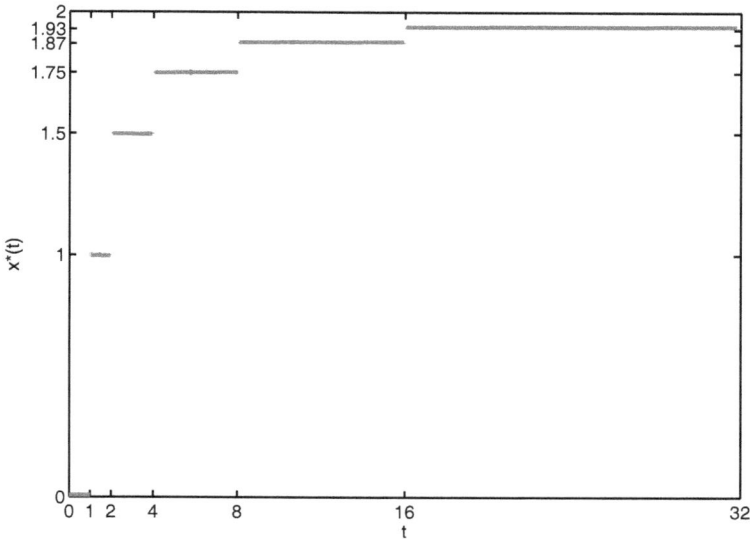

Fig. 5.1.   Solution $x^*(t)$ for $0 < t < +\infty$.

(w.r.t. t), i.e., the solution $x(t)$ should be constucted as $x(t) = \sum_k c_k \delta(t - T_k) + u(t)$, where regular part $u(t)$ (as well as in Example 5.3) is piecewise continuous function is constructed using the successive approximations and the method of steps.

PART 2

Nonlinear Models, Singularities and Control

Part Summary

In this part we study several classes of nonlinear models. In Chapter 5 we start with nonlinear Hammerstein integral equation (NHIE) with a single parameter and continue with proposed technique application for solution to the abstract operator equation with vector parameter. In Chapter 6 we study the Volterra integral equation with non-invertible operator in the leading part for the cases of continuous solution and for the case when equation possesses no continuous solution we design solution in the class of distributions. The methods from Chapter 5 are employed for solution of the nonlinear Volterra integral equation with non-invertible operator in here in Chapter 6. In Chapter 7 we describe an algorithm for construction the parametric families of small branching solutions of nonlinear differential equations of order n in the neighborhood of branching point and demonstrate it on the example of nonlinear differential equations appearing in magnetic insulation problem.

One of the most interesting control problem addressed in Chapters 8, 9, and 10. That problem appears in mathematical modelling and control of the nonlinear dynamical systems based on the Volterra series models and their generalization. In such models it is conventionally assumed that the transfer functions are already defined and it is necessary to determine the input signal which can deliver the desired and known *a priory* system response. In Chapter 8 we employ the convex majorants method to find the continuous solutions to such class of Volterra nonlinear equations. Chapters 9 and 10 provide the theory for generalized solutions construction to such nonlinear integral equations appearing in finite Volterra models. In Section 11.2 we overview the state-of-the-art Volterra models identification techniques.

Chapter 6

Nonlinear Hammerstein Integral Equations

Nonlinear Hammerstein integral equations (NHIE) appear in many problems in mathematical physics, the dynamic model of chemical reactor, certain problems in control theory, and various reformulations of an elliptic partial differential equation with nonlinear boundary conditions, this chapter deals with finding the analytical solutions of these problems and generalize the algorithm for the case of nonlinear operator equations with vector parameter in a non-regular case.

Outline

The aim of this chapter is the construction of successive approximations of branches of a solution to the Hammerstein integral equation with a single parameter $\lambda \in \mathbb{R}^1$ with explicit parametrization and the derivation of some iteration formulas convenient for numerical solution with a simple choice of an initial approximation. First of all, in Section 6.1 we formulate the problem, the state-of-the-art and the methods we employ for solution. Then in Section 6.2 we develop the techniques of successive approximations in a neighborhood of a bifurcation point with explicit parametrization of a solution while taking into account the peculiarities of nonlinear equations in question. The designed iteration scheme based on of the Lyapunov-Schmidt method, here readers may refer to the book of [Sidorov *et al.* (2002)]. In Section 6.3 we generalize the method for the case of abstract nonlinear operator equation with vector parameter λ from an open set $\Omega \subset \Lambda$, where Λ is a vector space with norm. We formulate and prove an existence Theorem. In Section 6.4 we apply the designed method for practical examples, including solution of the nonlinear boundary value problem which appears in modelling oscillations of a satellite in the plane of its elliptic orbit.

6.1   Problem Statement

In this section we consider the following equation

$$u(x) = \int_a^b K(x,s)g(s,u(s),\lambda)\,ds, \tag{6.1.1}$$

where $K(x,s), g(s,u,\lambda) = u(s) + f(s,u,\lambda)$ are continuous functions for $a \le x, s \le b, |u| \le r, |\lambda| \le \rho$,

$$f(s,u,\lambda) = \sum_{i=2}^{\infty} q_{io}(s)u^i + \sum_{i=0}^{\infty}\sum_{k=1}^{\infty} q_{ik}(s)u^i\lambda^k. \tag{6.1.2}$$

If unity is not a characteristic number of the kernel $K(x,s)$, then (6.1.1) possesses the unique solution such that $u \to 0$ as $\lambda \to 0$ and a broad class of methods can be used for its construction. Assume that unity is the characteristic number of the kernel $K(x,s)$ of rank n, while $\{\phi_i\}_1^n$ are the corresponding eigenfunctions and $\{\phi_i\}_1^n$ are the eigenfunctions of the adjoint kernel $K(s,x)$. We need to construct a solution in some neighborhood about the bifurcation point $\lambda = 0$ such that $u_\lambda \to 0$ as $\lambda \to 0$. The similar NHIE is studied in Lebesgue space in the paper of [O'Regan (2011)].

To construct the branching solutions to equation (6.1.1), we can employ the classical Trenogin's results of analytic bifurcation theory constructed in the books of [Trénoguine (1985); Vainberg and Trenogin (1974); Sidorov *et al.* (2002)], [Zeidler (1986)], ch.8 and [Buffoni and Toland (2003)]. Alongside asymptotic methods, of great importance in solving the equation (6.1.1) is the development of the techniques of successive approximations in a neighborhood of a bifurcation point which were not considered in the book of [Sidorov *et al.* (2002)]. The latter methods can be constructed on the base of explicit and implicit parametrization, in particular, under the condition of group symmetry of (6.1.1) (see results of [Moore (1980); Sidorov *et al.* (2002)]). For example, [Moore (1980); Sidorov *et al.* (2002); Keller (1977)] designed the schemes of successive approximations with implicit parametrization of branches of solutions to operator equations in Banach spaces. The implicit parametrization was used for solving particular problems also in the papers of [Sidorov (1997); Sidorov and Trenogin (1978)].

Meanwhile, for a qualitative analysis of branches of a solution, we often need to know the explicit dependence on some physically meaningful bifurcation parameter in the equation under study. Hence, it is of interest to develop the techniques of successive approximations in a neighborhood of a bifurcation point with explicit parametrization of a solution while taking into account the peculiarities of nonlinear equations in question. The aim of this chapter is the construction of successive approximations of branches of a solution to (6.1.1) with explicit parametrization and the derivation of some iteration formulas convenient for numerical solution of (6.1.1) with a simple choice of an initial approximation. Our iteration scheme relies on the methodology of the analytic Lyapunov-Schmidt method from the books of [Sidorov *et al.* (2002); Vainberg and Trenogin (1974)].

6.2 Asymptotics of a Solution and Successive Approximations

We will employ the following condition below

A. *The expansion (6.1.2) can be rearranged to the following form*

$$f(s, u, \lambda) = \sum_{i\nu+k=\theta}^{\infty} q_{ik}(s) u^i \lambda^k,$$

where $\nu = r/s, \theta = \frac{r+m}{s}$, while r, s, m are positive integers.

Note that in some particular cases r, s, and m can be calculated by the Newton polygon method (here readers may refer to textbook of [Trénoguine (1985)], pp.421–426). In this case it suffices to indicate the points (i, k) with integer coordinates corresponding to the nonzero coefficients q_{ik} on the coordinate plane and to construct the corresponding Newton diagram .

The unknowns ν are not chosen uniquely and are assumed to be equal to $\tan \gamma$, where γ is the angle between one of the segments of the diagram and the negative semi-axis. The respective θ is the ordinate of the point of intersection of the extension of this segment with the ordinate axis. Assume now that

B. *The system of algebraic equations*

$$l_j(c) \equiv \int_a^b \sum_{i\nu+k=\theta} q_{ik}(s)(c, \phi(s))^i \psi_j(s) \, ds = 0, \tag{6.2.1}$$

$j = 1, \ldots, n$, *where $(c, \phi) = \sum_{i=1}^n c_i \phi_i(s)$, has simple nonzero solution c^*.*

If kernel $K(x, s)$ is symmetric then **B** can be replaced with the following condition

B'. Function

$$U(c) = \int_a^b \sum_{i\nu+k=\theta} \frac{1}{i+1} q_{ik}(s)(c, \phi(s))^{i+1} \, ds \qquad (6.2.2)$$

has a non-degenerate critical point $c^* \neq 0$.

Theorem 6.1. *Let conditions* **A** *and* **B** *hold. Then (6.1.1) has a solution of the form*

$$u = \lambda^\nu(c^*, \phi(x)) + r(x, \lambda), \qquad (6.2.3)$$

where $|r(x, \lambda)| = o(|\lambda|^\nu)$ *as* $\lambda \to 0$*, and the function* $r(x, \lambda)$ *can be uniquely determined by the method of successive approximations.*

Proof. Introduce the operators

$$Bu = u - \int_a^b K(x, s)u(s) \, ds,$$

$$\hat{B}u = u - \int_a^b \hat{K}(x, s)u(s) \, ds, \quad R(u, \lambda) = \int_a^b K(x, s)f(s, u(s), \lambda) \, ds$$

where $\hat{K}(x, s) = K(x, s) - \sum_{i=1}^n \psi_i(x)\phi_i(s)$.

Observe that, by the Schmidt lemma (see the books of [Trénoguine (1985)], p.221, [Zeidler (1986)], p.130), the operator $\hat{B}$ has bounded inverse $\Gamma = I - \int_a^b R(x, s)[\cdot] ds$, where $R(x, s)$ is the resolvent of the kernel $\hat{K}(x, s)$.

We look for the a solution to (6.1.1) in the form

$$u = (\Gamma \omega(x) + (c, \phi(x)))\lambda^\nu, \qquad (6.2.4)$$

where $\omega(x)$ satisfies

$$(\omega, \psi_i) \equiv \int_a^b \omega(x)\psi_i(s) \, ds = 0, \quad i = 1, .., n. \qquad (6.2.5)$$

With change of variables (6.2.4) we rewrite equation (6.1.1) as follows

$$\omega = \lambda^{-\nu} R((\Gamma \omega + (c, \phi))\lambda^\nu, \lambda), \qquad (6.2.6)$$

We can add the following equalities to the last equation

$$\lambda^{-\theta}(R((\Gamma\omega + (c, \phi))\lambda^\nu, \lambda), \psi_i) = 0, \ i = 1, \ldots, n, \qquad (6.2.7)$$

which result from (6.2.5). Therefore, the problem is reduced to finding a function $\omega(x) \to 0$ and vector $c(\lambda) \to c^*$ as $\lambda \to 0$ from (6.2.6) and (6.2.7). It is to be noted that in (6.2.7)

$$(R(u, \lambda), \psi_i) = \int_a^b f(s, u, \lambda)\psi_i(s)\, ds, \ i = 1, .., n.$$

The system (6.2.6) and (6.2.7) can be regarded as the operator equation

$$\Phi(\boldsymbol{w}, \lambda) = 0, \qquad (6.2.8)$$

with the parameter λ. relative to $\boldsymbol{w} = (\omega, c)$ from $\mathcal{C}_{[a,b]} \oplus \mathbb{R}^n$. The nonlinear map Φ acts from the Banach space $E_1 = \mathcal{C}_{[a,b]} \oplus \mathbb{R}^{n+1}$ into the Banach space $E_2 = \mathcal{C}_{[a,b]} \oplus \mathbb{R}^n$. In accord with the choice of ν, r, s the operator Φ is continuous in some neighborhood about the point $\boldsymbol{w_0} = (0, c^*)$, $\lambda_0 = 0$, where c^* is nonzero solution to (6.2.1),

$$\lim_{\boldsymbol{w} \to \boldsymbol{w_0} \ \lambda \to 0} \Phi(\boldsymbol{w}, \lambda) = 0.$$

Furthermore, Φ has the continuous Fréchet derivative $\Phi_{\boldsymbol{w}}(\boldsymbol{w}, \lambda)$ with respect to $\boldsymbol{w}$ in some neighborhood about this point.

We note that

$$\frac{\partial}{\partial\omega}(R((\Gamma\omega + (c, \phi))\lambda^\nu, \lambda), \psi_i)$$

$$= \lambda^\nu \int_a^b f_u'((\Gamma\omega + (c, \Phi))\lambda^\nu, \lambda)\psi_i(s)\Gamma[\cdot]\, ds,$$

$$\frac{\partial}{\partial c_j}(R((\Gamma\omega + (c, \phi))\lambda^\nu, \lambda, \psi_i)) = \lambda^\nu \int_a^b f_u'(s, (\Gamma\omega + (c, \phi)\lambda^\nu, \lambda))\psi_i(s)\phi_j(s)\, ds,$$

$j = 1, \ldots, n$. The Fréchet derivative of Φ with respect to $\boldsymbol{w}$ at $(\boldsymbol{w_0}, 0)$ is representable as

$$\Phi_{\boldsymbol{w}}'(\boldsymbol{w_0}, 0) = \begin{pmatrix} I & 0 \\ L & \Xi \end{pmatrix},$$

where I is the identity operator from $\mathcal{C}_{[a,b]}$ in $\mathcal{C}_{[a,b]}$, while 0 is the zero operator from $\mathbb{R}^n$ in $\mathcal{C}_{[a,b]}$, L is linear functional from $\mathcal{C}_{[a,b]}$ in $\mathbb{R}^n$, $L = (L_1, \ldots, L_n)'$, where

$$L_i = \int_a^b \psi_i(s) f_u'(s, (c^*, \phi), 0) \Gamma(\cdot)\, ds$$

is linear functional from $\mathcal{C}_{[a,b]}$ in $\mathbb{R}^1$,

$$\Xi = \left[\frac{\partial l_i(c^*)}{\partial c_j} \right]_{i,j=1,\ldots,n}$$

is $n \times n$-matrix. The linear operator $\Phi_w'(w_0, 0)$ has bounded inverse since Ξ is non-degenerate matrix because of the condition **B**. Hence, (6.2.8) satisfies the conditions of the implicit function theorem (here readers may refer to the textbook of [Trénoguine (1985)], p.411) and the sought solution can be found by the successive approximations method as follows:

$$w_n = w_{n-1} - \Phi_w^{-1}(w_0, 0)\Phi(w_{n-1}, \lambda),\, w_0 = (0, c^*).$$

Thus, for determining a solution to (6.1.1), we need to construct the sequence

$$u_n = (\Gamma w_n + (c_n, \phi))\lambda^\nu,\, n = 1, 2, \ldots \qquad (6.2.9)$$

$$u_0 = (c^*, \phi)\lambda^\nu,$$

$w_0 = 0$, $c_0 = c^*$. Here $w_n = w_{n-1} + \Delta_n$, $c_n = c_{n-1} + \delta_n$. Function Δ_n and vector δ_n from $\mathbb{R}^n$ are defined by the formulas

$$\Delta_n(x) = \lambda^{-\nu} \int_a^b K(x, s) g(s, u_{n-1}(s), \lambda) ds - w_{n-1} \qquad (6.2.10)$$

$$\delta_n = -\lambda^{-\theta} \int_a^b f_u'(s, u_{n-1}, \lambda) \hat{\Psi}(s)\, ds$$

$$- \int_a^b f_u'(s, (c^*, \phi), 0) \Gamma \Delta_n \hat{\psi}(s) ds. \qquad (6.2.11)$$

The vector-function $\hat{\psi}(s) = (\hat{\psi}_1(s), \ldots, \hat{\psi}_n(s))'$ is determined as $\hat{\psi}(s) = \Xi^{-1}\psi(s)$. The claim is proven. $\qquad \square$

Remark 6.1. In order to construct $\Gamma\omega_n$, $\Gamma\Delta_n$ in (6.2.9) and (6.2.11), we can use the representation of Γ through the resolvent of $\hat{K}(x,s)$ or we can solve a linear Fredholm integral equation of the second kind with a kernel of this kind. If $K(x,s) = \sum\limits_{i=1}^{n} \phi_i(x)\phi_i(s)$, where ϕ_i is an orthonormal system, then Γ is the identity and successive approximations in (6.2.9)–(6.2.11) are reduced to quadratures.

Remark 6.2. By the choice of ν, r, s, the point $\lambda = 0$ is a removable singularity in the iteration formulas (6.2.10) and (6.2.11). If f is a polynomial in the unknown function u then we can "regularize" (6.2.10) and (6.2.11) by canceling the negative powers and inserting the representation of u_{n-1} in the form (6.2.4) in these formulas. Generally, in order to ensure the stability of calculations for small λ, we follow the idea of regularization in the Tikhonov sense, involve the results of [Bel'tukov and Shil'ko (1973)]. We can thus make the change of variables $\lambda^{-\nu} \Rightarrow (\lambda + \alpha(\delta))^{-\nu}$, $\lambda^{-\theta} \Rightarrow (\lambda + \alpha(\delta))^{-\theta}$ in (6.2.10), (6.2.11). In this case the regularizing parameter $\alpha(\delta)$ must agree with the error δ of calculations as suggested by [Bel'tukov and Shil'ko (1973)].

Remark 6.3. The analyticity of f can be replaced with the assumption that

$$\left| f(s,u,\lambda) - \sum_{\frac{i}{n_1}+\frac{k}{n_2}=1} q_{ik}(s)u^i\lambda^k \right| = o\left(\sum_{\frac{i}{n_1}+\frac{k}{n_2}=1} |u|^i|\lambda|^k \right)$$

as $u \to 0, \lambda \to 0$. In this case of $q_{n_10}(s) \neq 0$, $q_{0n_2}(s) \neq 0$, then the points of the Newton diagram of the function $f(s,u,\lambda)$ are not above the line $\frac{i}{n_1} + \frac{k}{n_2} = 1$.

Example 6.1. Consider the equation

$$u(x) - \frac{1}{\pi}\int_0^{2\pi} \cos x \cos s((1+\lambda)u(s) - su^3(s) + \lambda^3 \cos s)\,ds = 0.$$

The conditions of the Theorem and Remark 6.1 are satisfied. In this example we have in (6.2.4) $n = 1$, $\phi = \psi = \frac{\cos x}{\sqrt{\pi}}$, $\nu = 2$ or $\nu = \frac{1}{2}$. By Remark 6.1, Γ is an identity operator and the iteration formulas (6.2.10) and (6.2.11) of successive approximations are reduced to quadratures. There exist the following three branches of a solution with explicit parametrization:

$$u_1(x,\lambda) = -\lambda^2 \cos x + \mathcal{O}(|\lambda|^3),$$

$$u_{2,3}(x, \lambda) = \pm\sqrt{3\lambda}\frac{\pi}{2}\cos x + \mathcal{O}(|\lambda|).$$

To conclude this section we note that the above technique of successive approximations is applicable in the irregular case to solving nonlinear integral equations on using continuation in the length of the integration segment.

In the next section we generalize (see also the paper of [Sidorov *et al.* (2012)]) the above technique to the case of the operator equations in the Banach spaces.

6.3 Solution to the Abstract Operator Equation with Vector Parameter

Let X and Y be Banach spaces and let Λ be a normed vector space. We address the abstract nonlinear operator equation

$$Bu = F(u, \alpha(\lambda), \beta\lambda), \qquad (6.3.1)$$

where

$$F(u, \alpha, \beta) = F_N(u, \alpha, \beta) + \mathcal{O}((\|u\| + |\alpha| + |\beta|)^{N+1}),$$

$$F_N(u, \alpha, \beta) = \sum_{i+k+j=1}^{N} F_{ikj}(u)\alpha^k\beta^j,$$

$F_{100} = 0$, $F_{ikj}(u)$ are ith degree mapping from X into Y. The closed Fredholm operator B acts from X into Y and has the domain of definition dense in X. We assume that $\{\phi_i\}_1^n$ is a basis in $N(B)$ and $\{\psi_i\}_1^n$ is a basis in the defect space $N^*(B)$, $\alpha(\lambda)$ and $\beta(\lambda)$ are continuous functionals of the parameter λ from an open set $\Omega \subset \Lambda$, where Λ is a normed vector space, $0 \in \partial\Omega$, $\alpha(0) = 0$, and $\beta(0) = 0$. We call the set Ω a *sectorial neighborhood* of origin.

In this section we employ the similar with previous sections methodology to construct successive approximations to the continuous solutions $u(\lambda) \to 0$ as $\Omega \ni \lambda \to 0$ in a sectorial neighborhood Ω of $\lambda = 0$. In previous sections of this chapter we addressed solutions of various orders of smallness to the Hammerstein integral equation for the case of single parameter $\lambda \in \mathbb{R}^1$. We develop further the results of the Sections 6.1–6.3 and employ the analytical branching theory for solutions to operator equations presented in the books of [Trénoguine (1985); Vainberg and Trenogin (1974); Sidorov *et al.* (2002)].

An Existence Theorem

Suppose the fulfillment of
Condition A: There exist $\nu = r/s$ and $\theta = (r+m)/s$, where r, m, and s are positive integers such that the mapping F_N admits the representation

$$F_N(\alpha^\nu v, \alpha, 0) = \alpha^\theta \sum_{r\nu+k=\theta} F_{ik0}(v) + r(v, \alpha), \qquad (6.3.2)$$

$\|r(v, \alpha)\| = o(\alpha^\theta)$ and $\theta < N$.

In some specific cases, in order to find suitable r, m, and s, one has to draw the points (i, k) corresponding to the nonzero terms F_{ik0} in the expansion of the mapping $F_N(u, \alpha, 0)$ on the coordinate plane and construct the Newton diagram . The desired ν is put to be equal to $\tan\gamma$, where γ is the inclination angle of one of the segments of the diagram with the negative direction of the abscissa axis i. Here θ is equal to the ordinate of the point of intersection of the prolongation of this interval with the ordinate axis k. Since a Newton diagram can have several intervals, the choice of the r, m and s can be non-unique.

Let us suppose in a neighborhood of the origin, we have the following inequality

$$\|F(u, \alpha, \beta) - F(u, \alpha, 0)\| \leq L(\|u\|, |\alpha|, |\beta|)|\beta|,$$

where $L(\|u\|, |\alpha|, |\beta|) = \mathcal{O}(\|u\|, |\alpha|, |\beta|^l)$, $l \geq 0$. Suppose $\alpha(\lambda)$ and $\beta(\lambda)$ satisfy the following estimate in a sectorial neighborhood of Ω.
Condition B: $\beta(\lambda) = \bar{o}(\alpha^\theta(\lambda))$ for $\Omega \ni \lambda \to 0$.

Remark 6.4. If $l \geq \max\left(\frac{r+m}{s}, \frac{r+m}{r}\right)$ then Condition B can be replaced with the less restrictive condition $\beta = \bar{o}(\alpha^{\theta/(1+l)})$.

Moreover let us here suppose the fulfillment of
Condition C: the system of algebraic equations

$$L_j(c) \equiv \left\langle \sum_{ir+ks=r+m} F_{ik0}(c\phi), \psi_j \right\rangle = 0,$$

where $j = \overline{1, n}$ and $c\phi = \sum_{k=1}^{n} c_k \phi_k$, has a simple solution c^*. Introduce the Trenogin-Schmidt operator (for more details readers may refer to the textbooks of [Trénoguine (1985)], p.221, [Zeidler (1986)], p.376)

$$\Gamma = \left(B + \sum_{i=1}^{n} \langle \cdot, \gamma_j \rangle z_i\right)^{-1},$$

where $\langle \phi_i, \gamma_k \rangle = \delta_{ik}$ and $\langle z_i, \psi_k \rangle = \delta_{ik}$. For constructing small solutions to (6.3.1), let us employ the following uniformization

$$u = (\Gamma v(\lambda) + c(\lambda)\phi)\alpha(\lambda)^\nu, \qquad (6.3.3)$$

where $c(0) = c^*$ and $v(0) = 0$. The unknown function $u(\lambda)$ satisfies $\langle v, \psi_i \rangle = 0, i = \overline{1,n}$. Using the substitution (6.3.3) we reduce (6.3.1) to the following form

$$v = F((\Gamma v + c\phi)\alpha^\nu, \alpha, \beta)\alpha^{-\nu}, \qquad (6.3.4)$$

by complementing it with

$$\alpha^{-\theta} \langle F((\Gamma v + c\phi)\alpha^\nu, \alpha, \beta), \psi_j \rangle = 0, \; j = \overline{1,n}. \qquad (6.3.5)$$

Thus the problem of constructing the branches (6.3.3) is reduced to finding the functions $v(\lambda)$ and $c(\lambda) = (c_1(\lambda), \ldots, c_n(\lambda))'$ from (6.3.4) and (6.3.5). For $\Omega \ni \lambda \to 0$, search for a function $v(\lambda) \to 0$ and $c(\lambda) \to c^*$. Consider (6.3.4) and (6.3.5) as a single operator equation

$$\Phi(\omega, \lambda) = 0 \qquad (6.3.6)$$

with respect to $\omega = (v, c)$. Introduce the Banach space E of the elements w with the norm

$$w = \max_{\lambda \in \overline{\Omega}_1}(||v(\lambda)||_Y + |c(\lambda)|_{\mathbb{R}^n}),$$

where $\Omega_1 \subset \Omega \subset \Lambda$. The nonlinear mapping Φ, depending on a small vector parameter λ, acts from E into E. By the choice of the numbers $r, m,$ and s and the indicated asymptotic compatibility of the functionals $\alpha(\lambda)$ and $\beta(\lambda)$ in Ω (refer to Condition B), the operator Φ is continuous for $\lambda \in \Omega$ in a neighborhood of the point $\omega_0 = (0, c^*)$. Moreover,

$$\lim_{w \to w_0, \Omega \ni w \to 0} \Phi(w, \lambda) = 0.$$

Furthermore, $\Phi(w, \lambda)$ has a continuous Fréchet derivative with respect to w for $\lambda \in \Omega$ in a neighborhood of w_0 and

$$\Phi_w(w_0, 0) = \begin{pmatrix} I & 0 \\ l & \mathfrak{A} \end{pmatrix}.$$

Here I is the identity operator from Y into Y, O is the zero operator from $\mathbb{R}^n$ into Y, $l = (l_1, \ldots, l_n)'$,

$$l_j(\cdot) = \left\langle \sum_{ir+ks=r+m} F'_{ik}(c^*\phi)\Gamma(\cdot), \psi_j \right\rangle, \; j = \overline{1,n}$$

are linear functionals on Y,

$$\mathfrak{A} = \left\langle \sum ir + ks = r + mF'_{ik}(c^*\phi)\phi_l, \psi_j \right\rangle \Big|_{l,j=\overline{1,n}}$$

is non-degenerate $n \times n$ matrix. The linear bounded operator $\Phi_w(w_0, 0)$ acting from E into E has a bounded inverse:

$$\Phi_w^{-1}(w_0, 0) = \begin{pmatrix} I & 0 \\ -\mathfrak{A}^{-1}l(\cdot) & \mathfrak{A}^{-1} \end{pmatrix}.$$

Thus, for $\lambda \in \Omega$, in a neighborhood of w_0, equation (6.3.6) meets the conditions of Implicit Operator Theorem, here readers may refer to the textbooks of [Zeidler (1986)], p.150, [Trénoguine (1985)]. Consequently, there exists a domain $\Omega_1 \subset \Omega$, $0 \in \partial\Omega$, such that, the desired continuous solution $w \to w_0$ as $\Omega_1 \in \lambda \to 0$ can be found by the successive approximations

$$w_n = w_{n-1} - \Phi_w^{-1}(w_0, 0)\Phi(w_{n-1}, \lambda),$$

where $w_0 = (0, c^*)$, $n = 1, 2, \dots$. What has been said implies

Theorem 6.2. *Let Conditions A, B and C hold. Then there exists an open domain $\Omega_1 \subset \Omega, 0 \in \partial\Omega_1, \overline{\Omega}_1 \subset \overline{\Omega}$ such that if $\lambda \in \overline{\Omega}_1$ and $\beta(\lambda) = o(\alpha(\lambda))$ then (6.3.1) has a continuous solution:*

$$u(\lambda) = \alpha(\lambda)^\nu(c^*\phi + r(\lambda)), \tag{6.3.7}$$

where c^* is a simple root of system from Condition C, the function $r(\lambda)$ is defined uniquely by successive approximations, $\|r(\lambda)\| = o(1)$ as $\Omega \in \lambda \to 0$.

Corollary 6.1. *Under the conditions of the Theorem 6.2 with $\lambda \in \Omega_1$, the sequence $u_n = \hat{u}_n + c_n\phi\alpha^\nu$ converges to solution (6.3.7), where the functions $\hat{u}_n$ and c_n are calculated from the linear equations*

$$\check{B}\hat{u}_n = F(u_{n-1}, \alpha, \beta),$$

$$\mathfrak{A}(c_n - c_{n-1}) = -l(\check{B}(\hat{u}_n - \hat{u}_{n-1})) - \alpha^{-\theta} < F(u_{n-1}, \alpha, \beta), \psi >,$$

$$\hat{u}_0 = 0, \quad c_0 = c^*, \quad u_0 = c^*\phi\alpha^\nu, \ n = 1, 2, \dots.$$

Here $\check{B} = B + \sum_1^n < \cdot, \gamma_k > z_i$ is an invertible operator, $\mathfrak{A}$ is a nonnegative matrix,

$$l(\check{B}(\hat{u}_n - \hat{u}_{n-1})) = \left\langle \sum_{ir+ks=r+m} F'_{ik}(c^*\phi)(\hat{u}_n - \hat{u}_{n-1}), \psi \right\rangle$$

$$\psi = (\psi_1, \dots, \psi_n)'.$$

6.4   Examples

Example 6.2. The integral equation

$$u(t) - 3\int_0^1 tsu(s)\,ds = (\lambda_1^2 + \lambda_2^4)u(t) - u^3(t) + (\lambda_1^5 + \lambda_2^8)t$$

satisfies the hypothesis of the Theorem 6.2. Here $\psi = \phi = \sqrt{3}t$, $\alpha(\lambda) = \lambda_1^2 + \lambda_2^4$, $\beta(\lambda) = \lambda_1^5 + \lambda_2^8$, Ω is a punctured neighborhood of the zero in $\mathbb{R}^2$, $\nu = \frac{1}{2}$ and $\theta = \frac{3}{2}$. Obviously $\beta(\lambda) = o((\lambda_1^2 + \lambda_2^4)^{3/2})$; therefore, $u = (v + ct)\alpha(\lambda)^{1/2}$, where the functions $v(t)$ and $c(t)$ satisfy the following system

$$v(t) = (\lambda_1^2 + \lambda_2^4)(v + ct - (v + ct)^3) + \frac{\lambda_1^5 + \lambda_2^8}{\sqrt{\lambda_1^2 + \lambda_2^4}}t,$$

$$c - 3\int_0^1 (ct + v(t))^3 t\,dt + \frac{\lambda_1^5 + \lambda_2^6}{(\lambda_1^2 + \lambda_2^4)^{3/2}} = 0.$$

Following the conditions of the Theorem, we fix an initial approximation c_0 as one of the root of the equation $c - (3/5)c^3 = 0$. By the implicit function theorem, the obtained system has the three solutions as $\lambda \to 0$

$$(v_1, c_1) \to (0, \sqrt{5/3}), \quad (v_2, c_2) \to (0, -\sqrt{5/3}), (v_3, c_3) \to (0, 0)$$

which can be found with successive approximations. Consequently, the considered integral equation has three small solutions as $\lambda \to 0$:

$$u_{1,2} = \pm\sqrt{5/3(\lambda_1^2 + \lambda_2^4)}t + o\left(\sqrt{\lambda_1^2 + \lambda_2^4}\right), \quad u_3 = o\left(\sqrt{\lambda_1^2 + \lambda_2^4}\right)$$

Example 6.3. Let us now consider the following boundary value problem which model the oscillations of a satellite in the plane of its elliptic orbit (see Chapter 36 in the book of [Vainberg and Trenogin (1974)])

$$(1 + e\cos x)\frac{d^2u}{dx^2} - 2e\sin x\frac{du}{dx} + \alpha\sin u - 4e\sin x = 0 \qquad (6.4.1)$$

$$u(0) = u(\pi) = 0. \qquad (6.4.2)$$

Here e is the eccentricity and the parameter α depends on the principal central moment of inertia. [Vainberg and Trenogin (1974)] have constructed the branching solution of this problem and it is shown that problems (6.4.1)–(6.4.2) can have three real solutions. Based on the theory proposed in this section we can find asymptotic expansions between the

parameters e and α under which these solutions are constructed by successive approximations. Putting $\alpha = \alpha_0 + \lambda$, we can rewrite the problem as
an operator equation (6.3.1). In this case the linear operator $B \equiv \frac{d^2}{dx^2} + \alpha_0$
acts from $\mathring{C}^{(2)}_{[0,\pi]}$ into $C_{[0,\pi]}$ and is Fredholm; moreover, for $\alpha_0 = n^2$, we have
$\dim N(B) = 1, \phi(x) = \psi(x) = \sqrt{2/\pi}\sin x$. The nonlinear part of (6.3.1) is
as follows

$$F(u, \lambda, e) = -e\frac{d^2u}{dx^2} + 2e\frac{du}{dx} - \lambda\sin u + u - \sin u + 4e\sin x.$$

If $e = 0$ then, for every α, there exist a trivial solution $u_0 = 0$, and,
moreover, to the right of the critical (bifurcation) point $\alpha_0 = n^2$, the two
nontrivial solutions

$$u_{1,2} = \pm 2\sqrt{2}\sin x + o(\sqrt{\lambda})$$

bifurcate from the trivial solution. If e is small then, in a small neighborhood of the noncritical points α, there exists a single small solution.

Consider a more difficult case of constructing solutions for small $e > 0$
in the neighborhoods of the critical points $\alpha_0 = n^2$. Let for definiteness
$\alpha_0 = 1$. Then the cubic equation

$$\xi^3 - 2\pi^2\lambda\xi + 4\pi^3 e = 0 \qquad (6.4.3)$$

is the corresponding approximate branching equation. If $D = 27e^2 - 2\lambda^3 <$
0 then the Cardano formula guarantees that this equation has three real
solutions. Using the Theorem 6.2, one may prove the existence of the
corresponding real solutions to the boundary value and the possibility of
their construction by successive approximations. Here the Newton diagram
of the operator $F(u, \lambda, 0)$ consists of the segment passing through the points
$(3, 0)$ and $(1, 1)$. Therefore, in our case $r = 1, s = 2, m = 2$. Thus, based on
the Theorem 6.2, we must employ the following uniformization

$$u = \left(\Gamma v + c\sqrt{\frac{2}{\pi}}\sin x\right)\lambda^{1/2} \qquad (6.4.4)$$

with boundary conditions (6.4.2) under the asymptotic compatibility condition $e = o(\lambda^{3/2})$ on the parameters of the problem. Obviously, here
$D < 0$, i.e., the approximate branching equation has the three real roots.
The operator

$$\check{B} = \frac{d^2}{dx^2} + 1 + \frac{2}{\pi}\int_0^\pi \sin x \sin s[\cdot]\,ds \qquad (6.4.5)$$

with boundary conditions (6.4.2) has the bounded inverse $\Gamma \in \mathcal{C}_{[0,\pi]} \rightarrow \mathring{\mathcal{C}}^{(2)}_{[0,\pi]}$. By the Lagrange method of variation of arbitrary constants, taking into account the degeneracy of the kernel integral part of the operator $\check{B}$, we find the inverse operator $\Gamma v = \sin x * (I - P)v + pv - d\sin x$, where P is the projection and

$$P = \frac{2}{\pi} \int_0^\pi \sin x \sin s[\cdot]\, ds,$$

$$d = \frac{2}{\pi}(\sin x, \sin x * (I - P)v).$$

By Condition C, construct the algebraic equation $L(c) \equiv -c^3/(4\pi) + c = 0$ for defining an initial approximation of the coefficient c in (6.4.4). It has the three simple roots $\pm 2\sqrt{\pi}$ and 0. Therefore, by the Theorem 6.2, under the fulfillment of the asymptotic estimate $e = o(\lambda^{3/2})$, the boundary value problem has three real solutions

$$u_{1,2} = \pm 2\sqrt{2\lambda}\sin x + r_{1,2}(x, \lambda, \varepsilon),$$

$$u_3 = r_3(x, \lambda, \varepsilon),$$

where the functions r_i have the estimates $r_i(\lambda) = o(\lambda^{1/2})$ for $e = o(\lambda^{3/2})$. By the Corollary 6.1 we can refine the asymptotic approximations of the solutions $u_{1,2,3}$ by solving linear equations. If $e \sim c\lambda^n$ where $n < 3/2$, then the approximate solution to the branching equation (6.4.3) has a unique real solution since then obviously $D > 0$. In this case, the corresponding unique small real solution to the boundary value problem is as follows

$$u = -2(4e)^{1/3}\sin x + r(x, \lambda, e), \ ||r(x, \lambda, e)|| = o(e^{1/3}),$$

where, by the Corollary 6.1, $r(x, \lambda, e)$ is constructed uniquely by successive approximations.

Chapter 7

Nonlinear Volterra Operator Equations with Non-invertible Operator

7.1 Problem Statement

In this section we concentrate on the following nonlinear equation

$$G(u,t) + \int_0^t K(t,s,u(s))ds = 0, \ 0 \le s \le t \le \rho, \qquad (7.1.1)$$

where

$$G(u,t) = \sum_{i+k\ge 1}^{N} G_{ik}(u)t^k + R_1(u,t),$$

$$K(t,s,u) = \sum_{i+k+j\ge 0}^{N} K_{ikj}(u)t^k s^j + R_2(u,t,s).$$

The operators G and K act from the Banach space E_1 to the Banach space E_2 and are defined in a neighborhood of the point $u = 0, t = 0, s = 0$; $G_{ik}(u), K_{ikj}(u)$ are i-power operators. Operators R_1 and R_2 are continuous and Fréchet differentiable w.r.t. u and enjoy the following conditions

$$\|R_1(u,t)\| = o\left((\|u\| + |t|)^N\right),$$

$$\|R_2(u,t,s)\| = o\left((\|u\| + |t| + |s|)^N\right).$$

In this chapter we study the VIE with non-invertible operator in the main part. We study the case in which $G_{10} \equiv B$ is a closed Fredholm operator of index zero from E_1 to E_2 with domain dense in E_1. We assume that $\{\phi\}_1^n$ is a basis in $N(B)$, $\{\psi_i\}_1^n$ is a basis in the deficiency subspace $N^*(B)$. Equation (7.1.1) is of interest in applications, see for example the books of [Volmir (1967); Dolezal (1967)]. Number of initial-boundary value problems for PDE with non-invertible operator in the main part can be reduced to such equations. For more details the readers may refer to the books of [Sidorov et al. (2002); Sveshnikov et al. (2007); Sviridyuk and Fedorov (2003)]. The evolutionary character of the Volterra equations (with many practical problems) is discussed in the excellent book of [Pruss (2012)].

Outline

The aim of this chapter is to obtain sufficient conditions for the existence of solution of the VIE with algebraic branching points by developing the results of [Sidorov et al. (2002); Sidorov and Sidorov (2006); Falaleev (2000)]. We also employ the nonlinear analysis methods from the textbook of [Trénoguine (1985)], refer to its chapters 8 and 9 and from the book of [Zeidler (1986)].

In Section 7.2 we construct the asymptotic value of branches of solutions and indicate a successive approximation method uniformly converging in a neighborhood of the branching point. For the case in which the VIE (7.1.1) has no continuous solutions in Section 7.3 we suggest a method for constructing generalized solutions in the Sobolev-Schwartz distributions space. The general existence theorems discussed in this chapter are finally applied to the solution of an initial-boundary value problem.

7.2 Asymptotics in the Branching Points Neighborhood

We introduce the notation

$$G_N(u,t) = \sum_{i+k \geq 1}^{N} G_{ik}(u)t^k - G_{10}u.$$

Let the following condition be satisfied.
A. There exist $\nu = \frac{r}{s}, \theta = \frac{r+m}{s}$, were r, s, m are positive integers such that

$$G_N(t^\nu v, t) = t^\theta \sum_{i\nu+k=\theta} G_{ik}(v) + r(v,t) \qquad (7.2.1)$$

and the following estimates hold

$$||r(v,t)|| = o(t^\theta), \ ||R_1(t^\nu v, t)|| = o(t^\theta),$$

$$||K(t,t,t^\nu v)|| = \mathcal{O}(t^\theta).$$

In specific cases, one can readily compute the numbers r, s, and m by drawing the integer points (i, k) corresponding to nonzero terms $G_{ik}(u)$ on the coordinate plane and by constructing the corresponding Newton diagram [Trénoguine (1985)], Chapter 9, Item 37. The desired number ν is then taken to be $\tan \gamma$, where γ, is the angle between one segment of the diagram and the negative direction of the i-axis. The corresponding number θ is the k-intercept of the straight line containing this segment. Since the Newton diagram can contain several segments, it follows that the choice of ν and θ in the representation (7.2.1) may be non-unique. We assume that, in addition to Condition A, the following condition holds.

B. The system of algebraic equations

$$l_j(c) \equiv \left\langle \sum_{i\nu+k=\theta} G_{ik}(c\phi), \psi_j \right\rangle = 0, \tag{7.2.2}$$

where $j = 1, \ldots, n$, $c\phi = \sum_{1}^{n} c_i \phi_i$ has a simple nontrivial solution $c^* \neq 0$.

Following the book of [Trénoguine (1985)] we introduce the bounded Trenogin-Schmidt operator

$$\Gamma = \left(B + \sum_{i=1}^{n} \langle \cdot, \gamma_i \rangle z_i \right)^{-1},$$

where

$$\langle \phi_i, \gamma_k \rangle = \delta_{ik}, \quad \langle z_i, \psi_k \rangle = \delta_{ik}.$$

Here readers may also refer to [Zeidler (1986)], p.376. In this section we seek the solution of the VIE (7.1.1) in the form

$$u = (\Gamma v + c\phi)t^\nu. \tag{7.2.3}$$

The unknown function $v(t)$ satisfies the relations

$$\langle v, \psi_i \rangle = 0, \quad i = \overline{1, n}. \tag{7.2.4}$$

We rewrite (7.1.1) as follows

$$Bu = R(u, t), \tag{7.2.5}$$

where

$$R(u, t) = Bu - G(u, t) - \int_0^t K(u(s), t, s) \, ds.$$

Using the substitution (7.2.3), we reduce equation (7.2.5) to the form

$$v = t^{-\nu} R((\Gamma v + c\phi)t^\nu, t), \tag{7.2.6}$$

and supplement it with the following conditions

$$t^{-\theta}\langle R((\Gamma v + c\phi)t^{\nu}, t), \psi_j\rangle = 0, \qquad (7.2.7)$$

$j = 1, \ldots, n$, which follow from (7.2.4).

Therefore, the problem of constructing a solution $u(t)$ with an algebraic branching point for $t = 0$ is thereby reduced to finding functions $v(t)$ and $c = (c_1(t), \ldots, c_n(t))$ from system (7.2.6), (7.2.7). We seek $c(t) \to c^{\star}$ and $v(t) \to 0$ as $t \to 0$, where the vector $c^{\star}$ is a simple solution of system (7.2.2). In view of the representation (7.2.3) and the non-uniqueness of the solution $c^{\star}$, we call $t = 0$ an algebraic branching point of the solution of equation (7.1.1). Consider system (7.2.6), (7.2.7) as a single operator equation

$$\Phi(\omega, t) = 0, \qquad (7.2.8)$$

for the element $\omega = (v, c)$.

Let us introduce the Banach space X of elements ω with norm

$$||\omega|| = \max_{0 \le t \le \rho} (||v(t)||_{E_2} + |c(t)|_{R^n}).$$

Then the nonlinear mapping Φ acting from X to X. By virtue of the choice of ν, and θ operator Φ is continuous in a neighborhood of the point $\omega_0 = (0, c^{\star})$, $t_0 = 0$. In addition,

$$\lim_{\substack{\omega \to \omega_0 \\ t \to 0}} \Phi(\omega, t) = 0.$$

Moreover, the operator $\Phi(\omega, t)$ has continuous Fréchet derivative w.r.t. ω in the neighborhood of the point $\omega_0, t_0 = 0$

$$\Phi'_{\omega}(\omega, t)\Big|_{\omega=\omega_0, t=0} = \begin{pmatrix} I, & 0 \\ l, & \Xi \end{pmatrix}.$$

Here I is the identity mapping of E_2 into E_2, 0 is the zero operator from R^n to E_2, $l = (l_1, \ldots, l_n)'$,

$$l_j = \left\langle \sum_{i\nu+k=\theta} G'_{ik}(c^{\star}\phi)\Gamma\cdot, \psi_j \right\rangle, j = 1, \ldots, n,$$

are linear functionals defined on E_2,

$$\Xi = \left\langle \sum_{i\nu+k=\theta} G'_{ik}(c^{\star}\phi)\phi_s, \psi_j \right\rangle\Big|_{s,j=\overline{1,n}}$$

is a non-singular $n \times n$ matrix.

Let us introduce the system

$$\omega = \mu(\omega, t), \qquad (7.2.9)$$

where

$$\mu(\omega, t) = \omega - \big(\Phi_\omega(\omega_0, 0)\big)^{-1} \Phi(\omega, t).$$

By virtue of the above-introduced conditions, there exist radii $\rho > 0$, $r > 0$, positive constants m, l and continuous functions $q_1(\rho) \to 0$, $q_2(\rho) \to 0$ as $\rho \to 0$ such that

(1) $\|\Phi_\omega^{-1}(\omega_0, 0)\| \le m$;

(2) $\|\Phi_\omega(\omega_1, t) - \Phi_\omega(\omega_2, t)\| \le l\|\omega_1 - \omega_2\|$ as $\|\omega_i - \omega_0\| \le r$, $0 \le t \le \rho$;

(3) $m\|\Phi(\omega_0, t)\| \le q_2(\rho)$, $m\|\Phi_\omega(\omega_0, t) - \Phi_\omega(\omega_0, 0)\| \le q_1(\rho)$.

By choosing a sufficiently small $\rho > 0$ we ensure that the inequality

$$2ml < \frac{(1 - q_1(\rho))^2}{q_2(\rho)} \tag{7.2.10}$$

holds and set

$$r = \min\left\{ R, \frac{1 - q_1(\rho) - \sqrt{(1 - q_1(\rho))^2 - 2mlq_2(\rho)}}{ml} \right\}. \tag{7.2.11}$$

Then the operator μ takes to ball $s(\omega_0, r)$ into itself and is a contraction with ratio

$$1 - \sqrt{(1 - q_1(\rho))^2 - 2mlq_2(\rho)} < 1.$$

Indeed for $\|\omega - \omega_0\| \le r$ due to the above inequalities we have the chain of inequalities

$$\|\mu(\omega, t) - \omega_0\| = \left\| \Phi_\omega^{-1}(\omega_0, 0) \Big\{ \Phi_\omega(\omega_0, t)(\omega - \omega_0) - \Phi(\omega, t) \right.$$

$$\left. + \Phi(\omega_0, t) - \Phi(\omega_0, t) + (\Phi_\omega(\omega_0, 0) - \Phi_\omega(\omega_0, t))(\omega - \omega_0) \Big\} \right\|$$

$$\le m \left\| \int_0^1 (\Phi_\omega(\omega_0, t) - \Phi_\omega(\omega_0 + \Theta(\omega - \omega_0), t)) d\theta(\omega - \omega_0) \right\|$$

$$+ m\|\Phi(\omega_0, t)\| + m\|\Phi_\omega(\omega_0, 0) - \Phi_\omega(\omega_0, t)\| \|\omega - \omega_0\|$$

$$\le \frac{1}{2} mlr^2 + q_2(\rho) + q_1(\rho)r = r.$$

Therefore, by virtue of (7.2.11) and the choice of r under condition (7.2.10), the operator $\mu(\omega, t)$ is a mapping of $s(\omega_0, r)$ into itself.

Further, for $\omega \in S(\omega_0, r)$ we have the following estimate

$$\|\mu_\omega(\omega, t)\| \leq m \|\Phi_\omega(\omega_0, 0) - \Phi_\omega(\omega, t) + \Phi_\omega(\omega_0, t) - \Phi_\omega(\omega_0, t)\|$$

$$\leq m \|\Phi_\omega(\omega_0, 0) - \Phi_\omega(\omega_0, t)\| + m \|\Phi_\omega(\omega, t) - \Phi_\omega(\omega_0, t)\|$$

$$\leq q_1(\rho) + mlr \leq 1 - \sqrt{(1 - q_1(\rho))^2 - 2mlq_2(\rho)} < 1.$$

Consequently, the operator μ is a contraction operator with contraction ratio less than unity on the ball $S(\omega_0, r)$. Therefore, the desired solution $\omega(t) \to (0, c^*)$ as $t \to 0$ can be found by successive approximation method

$$\omega_n = \omega_{n-1} - \Phi_\omega^{-1}(\omega_0, 0)\Phi(\omega_{n-1}, t), \qquad (7.2.12)$$

where $\omega_0 = (0, c^*)$, $n = 1, 2, \ldots$

$$\Phi_\omega^{-1}(\omega_0, 0) = \begin{pmatrix} I, & 0 \\ -\Xi^{-1}l, & \Xi^{-1} \end{pmatrix}.$$

We have thereby proved the following theorem.

Theorem 7.1. *Let Conditions **A** and **B** be satisfied. Then there exists $\rho > 0$, such that for $0 \leq t \leq \rho$ equation (7.1.1) has a continuous solution*

$$u(t) = t^\nu(c^*\phi) + r(t), \qquad (7.2.13)$$

where $\|r(t)\| = o(t^\nu)$ as $t \to 0$. Function $r(t)$ is uniquely determined by the successive approximation method.

Remark 7.1. Theorem 7.1 can be refined by introducing the following conditions instead of the conditions **A** and **B**:
A′ :

$$G(t^\nu v, t) - G(0, t) = t^\nu P(v) + o(t^\nu),$$

where

$$G(0, t) = v_0 t^\nu + o(t^\nu)$$

as $t \to 0$.
B′ : The system

$$\langle P(\Gamma v_0 + c\phi), \psi_j \rangle = 0, \quad j = 1, .., n,$$

has a simple solution c^*. Then the result of Theorem 7.1 remains valid, but instead of (7.2.13) we have the following solution $u = t^\nu(\Gamma v_0 + c^*\phi) + r(t)$, where $\|r(t)\| = o(t^\nu)$. Recall that $v_0(x) \equiv 0$ in the assumptions of Theorem

7.1. Therefore, Remark 7.1 refines the Theorem 7.1.

The result of theorem also remains valid for the perturbed equation

$$\widetilde{G}(u,t) + \int_0^t \widetilde{K}(t,s,u(s))ds = 0,$$

provided that the perturbed operators $\widetilde{G}, \widetilde{K}$ are continuous and differentiable with respect to u and satisfy the asymptotic estimates

$$\|\widetilde{G}(u,t) - G(u,t)\| = o(t^\nu),$$

$$\|\widetilde{K}(t,t,t^\nu v)\| = \mathcal{O}(t^\theta),$$

$$\|Q(\widetilde{G}(u,t)) - G(u,t))\| = o(t^\theta),$$

where $Q = \sum_{i=1}^n <\cdot, \psi_i> z_i$ is a projector in E_2.

7.3 Perturbation of Linear Equation by Nonlinear Volterra Operator

Let us consider the equation

$$Bu = b(t) + \int_0^t K(t,s,u(s))ds, \quad 0 \le t \le \rho, \tag{7.3.1}$$

where B is the Fredholm operator, $b(t)$ is a differentiable function ranging in E_2, and the mapping $K(t,s,u)$ is differentiable w.r.t. t and continuous w.r.t. s and satisfies the Lipschitz condition w.r.t. u. The class of initial-boundary value problems studied in the books of [Sveshnikov *et al.* (2007); Sviridyuk and Fedorov (2003); Sidorov *et al.* (2002)] can be reduced to the equation (7.3.1). First we consider the case of $b(0) \in R(B)$ and the equation (7.3.1) may possess continuous solutions.

We introduce the following condition

C. The system

$$l_j(c) \equiv \langle b'(0) + K(0,0,\Gamma b(0) + c\phi), \psi_j \rangle = 0, j = 1, \ldots, n, \tag{7.3.2}$$

has a simple solution c^*.

Theorem 7.2. *Let the Condition* **C** *be satisfied, and let* $b(0) \in R(B)$. *Then equation (7.3.1) possesses a continuous solution*

$$u = \Gamma b(0) + c^*\phi + r(t),$$

where $r(t) \to 0$ *as* $t \to 0$, *and* $r(t)$ *is uniquely determined.*

Proof. Let's employ the substitution $u = \Gamma v(t) + c(t)\phi$, where $\langle v(t), \psi_j \rangle = 0, j = 1, \ldots, n$, we reduce the VIE (7.1.1) to the following system

$$v(t) = b(t) + \int_0^t K(t, s, \Gamma v(s) + c(s)\phi)ds, \qquad (7.3.3)$$

$$\langle b(t) + \int_0^t K(t, s, \Gamma v(s) + c(s)\phi)ds, \psi_j \rangle = 0, \qquad (7.3.4)$$

$j = 1, \ldots, n.$

Since $b(0) \in R(B)$, it follows that relation (7.3.4) can be replaced by the equations

$$\langle K(t, t, \Gamma v(t) + c(t)\phi) + b'(t) + \int_0^t K_t(t, s, \Gamma v(s) + c(s)\phi))ds, \psi_j \rangle = 0, \quad (7.3.5)$$

$j = 1, \ldots, n.$

Let us now consider the system (7.3.3), (7.3.5) in the neighborhood of the point $\omega_0 = (b(0), c^*)$, $t_0 = 0$ as single operator equation $\Phi(\omega, t) = 0$. As result, the operator

$$\Phi_\omega(\omega_0, 0) = \begin{bmatrix} I, & 0 \\ \langle K_u(0, 0, \Gamma b(0) + c^*\phi)\Gamma \cdot, \Psi \rangle, & \Xi \end{bmatrix},$$

where $\Psi = (\psi_1, \ldots, \psi_n)'$, $\Xi = [\langle K_u(0, 0, \Gamma b(0) + c^*\phi)\phi_i, \psi_j r \rangle]_{i,j=1,..,n}$ is a non-singular matrix, has a bounded inverse operator. Therefore, it follows from the proof of Theorem 7.1 that the sequence

$$\omega_n = \omega_{n-1} - \Phi_\omega^{-1}(\omega_0, 0)\Phi(\omega_{n-1}, t), \ \omega_0 = (b(0), c^*),$$

converges in a neighborhood of the point $t = 0$ to the desired solution. The theorem is proved.   □

Now let $b(0) \notin R(B.)$ In this case, equation (7.3.1) possesses no continuous solution. Following the papers of [Volmir (1967); Sidorov and Sidorov (2011a, 2006); Sidorov *et al.* (2010c); Sidorov (2011a)], we seek a solution in the class of distributions. To this end, consider the Banach space E as a subset of linear continuous functionals in E^{**} defined on E^*. Let us introduce the following definitions.

Definition 7.1. The set of infinitely differentiable functions $s(t)$ compactly supported on the interval $(0, \rho)$ and ranging in E^* is denoted by $(D; E^*)$. The set of linear continuous functionals defined on $(D; E^*)$ and ranging in E is referred to as the space of distributions and is denoted by $(D; E)$.

Definition 7.2. An element u in $(D'; E_1)$ is called a *generalized solution* of the equation $F(u, t) = 0$, where $F : E_1 \to E_2$, if identity

$$\langle F(u, t), s(t) \rangle = 0$$

holds for $\forall s(t) \in (D; E_2^*)$.

We assume that the following condition is satisfied
D.

$$K(t, s, u) = Ku + g(t, s, u),$$

where

$$K \in \mathcal{L}(E_1 \to E_2),$$

$$g(t, s, u) = \sum_{k,j=0}^{\infty} \sum_{i=1}^{\infty} q_{ikj}(u) t^k s^{i+j},$$

$q_{ikj}(u)$ are i-power operators from E_1 in E_2,

$$\det[\langle K\phi_i, \psi_j \rangle]_{i,j=\overline{1,n}} \neq 0.$$

In the class $(D'; E_1)$ we seek a generalized solution to the equation (7.3.1) in the form

$$u = (a\phi)\delta(t) + \Gamma v(t) + c(t)\phi, \tag{7.3.6}$$

where $a\phi = \sum_1^n a_i\phi_i$, $a \in R^n$, $\delta(t)$ is the Dirac delta function, $v(t)$ is regular function of t ranging in E_2, $c(t) = (c_1(t), \ldots, c_n(t))'$ is regular vector function,

$$\langle v(t), \psi_j \rangle = 0, \quad j = \overline{1, n}. \tag{7.3.7}$$

Since (here readers may refer to the books of [Kanwal (2013)], [Vladimirov (1979)]) $s\delta(s) = 0$, it follows that, by substituting (7.3.6) into (7.3.1) and taking into account the relation $B\Gamma v = v$, which holds by virtue of (7.3.7) we obtain the equation

$$v(t) = b(t) + Ka\phi + \int_0^t \left\{ K(\Gamma v(s) + c(s)\phi) + g(t, s, \Gamma v(s) + c(s)\phi) \right\} ds \tag{7.3.8}$$

with the conditions

$$\left\langle b(t) + Ka\phi + \int_0^t \{ K(\Gamma v(s) + c(s)\phi) + q(t, s, \Gamma v(s) + c(s)\phi) \} ds, \psi_j \right\rangle$$
$$= 0, \quad j = \overline{1, n}. \tag{7.3.9}$$

We choose a vector a^* as a solution of the following system of linear algebraic equations

$$\langle b(0) + Ka\phi, \psi_j \rangle = 0, \ j = \overline{1, n}.$$

Then conditions (7.3.9) become equivalent to the equations

$$\langle b'(t) + K\Gamma v(t) + Kc(t)\phi + g(t, t, \Gamma v(t) + c(t)\phi)$$

$$+ \int_0^t q_t(t, s, \Gamma v(s) + c(s)\phi)ds, \psi_j \rangle = 0, \qquad (7.3.10)$$

$j = 1, \ldots, n.$

Consider the system (7.3.8)–(7.3.10) as the single operator equation $\Phi(\omega, t) = 0$ in the neighborhood of the point $\omega_0 = (v_0, c^*)$, $t_0 = 0$, where $v_0 = b(0) + Ka^*\phi$, and c^* is the unique solution to the system of linear algebraic equations

$$\langle Kc\phi, \psi_j \rangle + \langle b'(0) + K\Gamma v_0, \psi_j \rangle = 0, j = \overline{1, n}.$$

In addition the operator

$$\Phi_\omega(\omega_0, 0) = \begin{pmatrix} I, & 0 \\ \langle K\Gamma \cdot, \Psi \rangle, & \Xi \end{pmatrix},$$

where $\Xi = [\langle K\phi_i, \psi_j \rangle]_{i,j=\overline{1,n}}$ is non-degenerate matrix, has a bounded inverse.

Consequently, by the proof of Theorem 7.2, there exists an element $\omega \to (v_0, c^*)$ as $t \to 0$ and it is uniquely determined by the successive approximation method. We have thereby proved the following assertion.

Theorem 7.3. *Let the condition D be satisfied and let* $b(0) \notin R(B)$. *The equation (7.3.1) possesses the following solution in the class* $(D'; E_1)$

$$u = a^*\phi\delta(t) + \Gamma(b(0) + Ka^*\phi) + c^*\phi + r(t),$$

where a^* and c^* are some vectors from $\mathbb{R}^n$, function $r(t)$ is uniquely defined, and $r(t) \to 0$ as $t \to 0$.

Example 7.1. Consider the following equation

$$\frac{\partial}{\partial t}\left(\frac{\partial^2 u}{\partial x^2} + u\right) + R\left(u, \frac{\partial u}{\partial x}, \frac{\partial^2 u}{\partial x^2}, t, x\right) = 0, \qquad (7.3.11)$$

which is a model equation, here readers may refer to [Sveshnikov *et al.* (2007)], p.94. We introduce the initial-boundary conditions

$$u|_{x=0} = 0, u|_{x=\pi} = 0, \qquad (7.3.12)$$

$$\left(\frac{\partial^2 u}{\partial x^2} + u\right)\Bigg|_{t=0} = v_0(x). \qquad (7.3.13)$$

Problems (7.3.11)–(7.3.13) can be reduced to the equation

$$\frac{\partial^2 u}{\partial x^2} + u - v_0(x) + \int_0^t R\left(u(t,x), \frac{\partial u(t,x)}{\partial x}, \frac{\partial^2 u(t,x)}{\partial x^2}, t, x\right) dt = 0, \quad (7.3.14)$$

with the boundary conditions (7.3.12). By the notation introduced above, in this equation, the operator $B = \frac{\partial^2}{\partial x^2} + 1$ is a mapping of $\overset{o}{\mathcal{C}}\,^{(2)}_{[0,\pi]}$ into $\mathcal{C}_{[0,\pi]}$ and its a Fredholm operator. In addition $dimN(B) = 1$, $\phi = \psi = \sin x$,

$$\Gamma v = \frac{2}{\pi} \int_0^\pi \sum_{k=2}^\infty \frac{\sin kx \sin ks}{1-k^2} v(s)\, ds.$$

We consider the following two cases:

Case 1.

The function $v_0(x)$ is coordinated with the boundary condition (7.3.12), i.e., $\int_0^\pi v_0(x)\sin x\, dx = 0$. Then $v_0 \in R(B)$ and the Theorem 7.2 can be employed. In accordance with Condition C (refer to the Theorem 7.2) to this end, we introduce the algebraic equation

$$\int_0^\pi \sin x R\big((\Gamma v_0)(x) + c\sin x, (\Gamma v_0)^{(1)}_{(x)}$$

$$+ c\cos x, (\Gamma v_0)^{(2)}_{(x)} - c\sin x, 0, x\big) dx = 0, \qquad (7.3.15)$$

under the assumption that the function $R(u, \frac{\partial u}{\partial x}, \frac{\partial^2 u}{\partial x^2}, t, x)$ satisfies the Lipschitz condition w.r.t. u, u_x, u_{xx} and differentiable w.r.t. t. The by the Theorem 7.2, each simple root c^* of the equation (7.3.15) corresponds to a solution of problems (7.3.11)–(7.3.13) of the form

$$u = (\Gamma v_0)(x) + c^* \sin x + r(t,x),$$

where function $r(t,x)$ is uniquely defined and $r(t,x) \to 0$ as $t \to 0$.

Case 2.

Let $v_0 = 0$. In addition there exist positive rational values $\nu = \frac{r}{s}$, $\theta = \frac{r+m}{s}$ such that

$$R\left(t^\nu v, t^\nu \frac{\partial v}{\partial x}, t^\nu \frac{\partial^2 v}{\partial x^2}, t, x\right) = t^\theta P\left(v, \frac{\partial v}{\partial x}, \frac{\partial^2 v}{\partial x^2}, x\right) + o(t^\theta)$$

and the algebraic equation

$$\int_0^\pi P(c\sin x, c\cos x, -c\sin x, x)\sin x\, dx = 0$$

has a simple solution $c^* \neq 0$, then problems (7.3.11)–(7.3.13) possesses the solution

$$u = t^\nu c^* \sin x + o(t^\nu).$$

The case in which the function $v_0(x)$ is not coordinated with the boundary conditions (7.3.12) requires a separate investigation. In this case, by Theorem 7.3, under certain conditions on the function R, the solution can be constructed in the class of distributions $(D'; \overset{o}{C}{}^{(2)}_{[0,\pi]})$. Finally, just as in Oskolkov-Benjamin-Bona-Mahony equations (reader may refer to the book of [Sveshnikov *et al.* (2007)], p.94 for more details), to be definite, we set

$$R(u, u_x, u_{xx}, t, x) = \frac{\partial u}{\partial x} + u\frac{\partial u}{\partial x} + \frac{\partial^2 u}{\partial x^2}.$$

Then the algebraic equation (7.3.15) is a linear equation of the following form

$$\left(-\frac{\pi}{2} + \frac{1}{6}\int_0^\pi \sin 2s v_0(s)ds\right)c + b = 0,$$

where $b = \frac{1}{2}\int_0^\pi \frac{d}{dx}(\Gamma v_0)^2 \sin x\, dx$. If in addition $\int_0^\pi \sin 2s v_0(s)ds \neq 3\pi$, then by the Theorem 7.2, there exists the solution $u = -b[-\frac{\pi}{2} + \frac{1}{6}\int_0^\pi \sin 2s v_0(s)ds]^{-1}\sin x + \Gamma v_0 + r(t, x)$, where $r(t, x) \to 0$ as $t \to 0$.

Remark 7.2. Of particular interest is the study of the problem (7.3.1), where operator B is not Fredholm operator but meets conditions of the paper of [Kiselev and Faddeev (2000)].

Chapter 8

Nonlinear Differential Equations Near Branching Points

In this chapter we describe an algorithm for construction the parametric families of small branching solutions of nonlinear differential equations of order n in the neighborhood of branching points. We propose new technique based on the methods of analytical theory of branching solutions of nonlinear equations and the theory of differential equations with singular point of the first kind. We illustrate the general existence theorems on example of nonlinear differential equations appearing in magnetic insulation problem.

Outline

This chapter is organized as follows. In Section 8.1 we provide the problem statement and outline the state-of-the-art in the field. In Section 8.2 the method for construction of the principal term of the solution is described. Also in this section we reduce the problem to a system of differential equations with a regular singular point w.r.t. the parametric family of solutions with the special structure depending on behavior of the roots of characteristic polynomial. In Section 8.2.1 we consider the technique for constructing the asymptotics of parametric families of solutions of the equation obtained in Section 8.2 and adduce the existence theorems for small solution of the main equation. In Section 8.3 we demonstrate this method's application for magnetic insulation problem.

8.1 Problem Statement

Consider a continuous function $F(x_n, x_{n-1}, \ldots, x_1, x, t)$ defined in a neighborhood of origin. Here $F = \sum\limits_{|i|+k=1}^{N} F_{ik}(\bar{x}) t^k + R(x, t)$,

$$F_{ik}(\bar{x}) = F_{i_n, i_{n-1}, \ldots, i_1, i_0} x_n^{i_n} .. x_1^{i_1} x^{i_0}, \ |i| = i_n + i_{n-1} + \cdots + i_1 + i_0.$$

The function $R(x, t)$ enjoys the bound $|R(x, t)| = o((|x| + |t|)^N)$. The main equation to be addressed in the chapter is as follows

$$F(x^{(n)}(t), x^{(n-1)}(t), \ldots, x^{(1)}(t), x(t), t) = 0, \ 0 \le t \le p. \qquad (8.1.1)$$

Our objective here is to find a solution to the equation (8.1.1) such as for $t \to 0 \ t^i x^{(i)}(t) \to 0$, for $i = 0, 1, \ldots, n$.

Definition 8.1. If a solution $x(t)$ is representable in the form

$$x = t^\varepsilon (x_0 + v(t)), \qquad (8.1.2)$$

where $\varepsilon = \frac{r}{s}$ is a rational positive number, $x_0 \ne 0$, ε, x_0 are defined non-uniquely, the function $v(t)$ tends to zero as $t \to 0$ and possibly depends on free parameters, $t^i x^{(i)}(t) \to 0$ as $t \to 0$, $i = \overline{0, n}$, then we call $t = 0$ the *branching* point of the small solution to equation (8.1.1).

Classes of differential equations (ordinary and partial ones) which are not resolved with respect to major derivatives have been studied in recent years by many mathematicians. See, for example, monographs [Trénoguine (1985)] and [Vainberg and Trenogin (1974); Sidorov *et al.* (2002)] for various approaches and related references. In many applications (see, for example, [Abdallah *et al.* (1997)]) it is necessary to construct solutions near branching points.

   Our objective is to find small solutions to (8.1.1) on the base of the analytical branching theory (readers may refer here to classical textbooks of [Trénoguine (1985)], Chapter 9 and [Zeidler (1986)]) and the theory of differential equations with a regular singular point (see [Vainberg and Trenogin (1974)] or [Coddington and Levinson (1955)]).

   For calculating the principle term $t^\varepsilon x_0$ of solution (8.1.2) we use the method of Newton diagram (here readers may refer for example to the books of [Vainberg and Trenogin (1974)], [Trénoguine (1985)], Chapter 9, [Bryuno (2000)]). The determination of the function $v(t)$ in solution (8.1.2) is reduced to the study of nonlinear differential equations in the form

$$\mathcal{L}\left(t\frac{d}{dt}\right) v = M(t^{n-1} v^{(n-1)}, \ldots, t v^{(1)}, v, t^{1/s}), \qquad (8.1.3)$$

where $\mathcal{L} = t^n \frac{d^n}{dt^n} + a_{n-1}(\varepsilon)t^{n-1}\frac{d^{n-1}}{dt^{n-1}} + \cdots + a_0(\varepsilon)$ is the Euler differential operator. We construct c-parametric families of solutions $v(t,c) \to 0$ as $t \to 0$ to the equation (8.1.3). The structure of the family $v(t,c)$ depends on roots of the characteristic polynomial

$$L(\lambda) = \lambda(\lambda - 1) \ldots (\lambda - (n-1)) + a_{n-1}(\varepsilon)\lambda(\lambda - 1)$$
$$\cdots (\lambda - (n-2)) + \cdots + a_0(\varepsilon) \tag{8.1.4}$$

of the Euler differential operator and on the form of the Jordan normal form of the matrix A expressed in terms of coefficients of the Euler operator. On this base we propose a technique for constructing the asymptotics of the function v as logarithmic-power sums of the Fuchs–Frobenius type (for more details readers may refer to classical book of [Coddington and Levinson (1955)], Chapter 4). Then we use the asymptotics as the initial approximation in the method of successive approximations. In the analytical case we expand the corresponding solution (8.1.2) in powers of $t^{1/s}$, $t \ln t$, $t^{\Re e \lambda_i} \cos(\Im m \lambda_i \ln t)$, $t^{\Re e \lambda_i} \sin(\Im m \lambda_i \ln t)$, where λ_i are roots of the characteristic polynomial $L(\lambda)$, $\Re e \lambda_i > 0$.

8.2   The Principal Term of the Solution. Problem Reduction

Introduce the denotation $F^N(\bar{x}, t) = \sum\limits_{|i|+k=1}^{N} F_{ik}(\bar{x})t^k$ and two conditions:

A. There exist rational numbers $\varepsilon = \frac{r}{s} > 0$, and $\theta = \frac{r_1}{s_1}$ such that the expansion of F^N can be reduced to the form

$$F^N(\bar{x}, t) = \sum_{\varepsilon|i|+k-i_1-2i_2-\cdots-ni_n \geq \theta} F_{i_k}(\bar{x})t^k.$$

B. $R(t^\varepsilon x, t) = o(t^\theta)$ with any x in a neighborhood of origin.

In concrete cases one can easily calculate ε, θ by plotting integer points $(|i|, k-i_1-2i_2-\cdots-ni_n)$, that correspond to nonzero $|i|$-homogeneous forms $F_{ik}(\bar{x})$ on the coordinate plane and by constructing the Newton diagram for these points. We set the desired value of ε to $\tan \phi$, where ϕ is the angle of the inclination of one of segments of the diagram to the negative direction of the abscissa. The corresponding value of θ equals the ordinate of the point of the intersection of the continuation of this segment with the coordinate axis. θ can be negative. In the latter case condition **B** is satisfied automatically due to the bound $R(x,t) = [(|x| + |t|)^N]$. Evidently, condition **B** is also satisfied with any positive θ, if the Newton diagram is located below the straight line that goes through points $(0, N)$ and $(N, 0)$. Since Newton diagram can have several segments, the choice of ε and θ can be non-unique.

Let us introduce the Euler differential operators of the kth order

$$l_k(u) = t^k u^{(k)} + c_k^1 \varepsilon t^{k-1} u^{(k-1)} + c_k^2 \varepsilon(\varepsilon - 1) t^{k-2} u^{(k-2)}$$

$$+ \cdots + \varepsilon(\varepsilon - 1) \ldots (\varepsilon - (k-1)) u, \; k = \overline{1, n}.$$

It is to be noted that $(t^\varepsilon u)^{(k)} = t^{\varepsilon - k} l_k(u)$. Then in view of condition **A** by changing the variable $x(t) = t^\varepsilon u(t)$ we obtain the expansion

$$F(x^{(n)}, \ldots, x^{(1)}, x, t) = t^\theta \sum_{\varepsilon|i| + k - i_1 - 2i_2 - \cdots - n i_n = \theta} F_{ik}(l_n(u), \ldots, l_0(u))$$

$$+ r(t^n u^{(n)}, \ldots, t u^{(1)}, u, t).$$

Here for the sake of symmetry we denote $l_0(u) = u$. Due to the choice of numbers ε and θ we have the following estimate

$$|r(t^n u^{(n)}, \ldots, t u^{(1)}, u, t)| = o(t^\theta). \tag{8.2.1}$$

In order to determine the coefficient x_0 in the desired solution (8.1.2) we introduce the polynomial

$$Q(l_n(x), \ldots, l_0(x)) = \sum_{\varepsilon|i| + k - i_1 - 2i_2 - \cdots - n i_n = \theta} F_{ik}(l_n(x), \ldots, l_0(x)),$$

where $l_k(x) = \varepsilon(\varepsilon - 1) \ldots (\varepsilon - (k-1)) x, \; k = \overline{1, n}$. We assume that apart from the conditions **A** and **B** the following condition is also satisfied

C. The polynomial $Q(l_n(x), \ldots, l_0(x))$ has a root $x_0 \neq 0$, and $\left. \frac{\partial Q}{\partial l_n(x)} \right|_{x=x_0} \neq 0$.

Then by changing the variable by formula (8.1.2) and dividing (8.1.1) by t^θ we reduce it to the following equation with respect to the function $v(t)$:

$$\mathcal{A}_n l_n(v) + \cdots + \mathcal{A}_0 l_0(v) + P(t^n v^{(n)}, \ldots, t v^{(1)}, v, t^{1/s}) = 0. \tag{8.2.2}$$

Here $\mathcal{A}_i = \left. \frac{\partial Q}{\partial l_i} \right|_{x=x_0}, \; i = \overline{0, n}, \; |P| = o(1)$ as $t \to 0$, $l_i(v)$ are the Euler differential operators.

Since $\mathcal{A}_n \neq 0$, by substituting differential operators $l_i(v)$ in (8.2.2) we obtain the equation

$$t^n v^{(n)}(t) + a_{n-1}(\varepsilon)t^{n-1}v^{(n-1)}(t) + \cdots + a_0(\varepsilon)v(t)$$
$$+ \mathcal{A}_n^{-1}P(t^n v^{(n)}(t), \ldots, tv^{(1)}, v(t), t^{1/s}) = 0, \qquad (8.2.3)$$

where $a_{n-1}(\varepsilon) = c_n^1 \varepsilon + \mathcal{A}_n^{-1}\mathcal{A}_{n-1}$, $a_{n-k}(\varepsilon)$, $k = 2, \ldots, n$ are certain k-order polynomials of ε.

Thus, in order to determine functions $v(t)$ we obtain equation (8.2.3) with Euler operator $\mathcal{L}(t\frac{d}{dt})v$ of nth order in the main part. It is to be noted that the characteristic polynomial $L(\lambda)$ in (8.1.4) is the characteristic polynomial of just this Euler operator. Since $|P| = o(1)$ as $t \to 0$ with any v, equation (8.2.3) allows one to use the method of successive approximations for determining the element $t^n v^{(n)}$ in the neighborhood of origin as function of $v(t), tv^{(1)}, \ldots, t^{n-1}v^{(n-1)}, t^{1/s}$.

As a result, the problem on determining the function v in our statement is reduced to an equation in the form (8.1.3).

Note that in (8.1.3) the function M and its first derivatives in $v, tv^{(1)}, \cdots, t^{n-1}v^{(n-1)}$ vanish at zero. The replacement $v = v_1, tv^{(1)} = v_2, \ldots, t^{n-1}v^{(n-1)} = v_n$ turns equation (8.1.3) into the following system of n nonlinear differential equations with a singular point of the first kind at $t = 0$:

$$t\frac{d\bar{v}}{dt} = A\bar{v} + f(\bar{v}, t^{1/s}), \qquad (8.2.4)$$

where $\bar{v} = (v_1, \ldots, v_n)'$,

$$A = \begin{vmatrix} 0 & 1 & 0 & \cdots & & 0 \\ 0 & 1 & 1 & \cdots & \cdot & 0 \\ 0 & 0 & 2 & \cdots & \cdot & 0 \\ \vdots & \vdots & \vdots & \vdots & \vdots & \vdots \\ 0 & 0 & 0 & \cdots & n-2 & 1 \\ -a_0(\varepsilon) & -a_1(\varepsilon) & -a_2(\varepsilon) & \cdots & -a_{n-2}(\varepsilon)) & n-1-a_{n-1}(\varepsilon)) \end{vmatrix},$$

$f(0, \ldots, 0, M(v_n, v_{n-1}, ..v_1, t^{1/s}))'$, $M(0, \ldots, 0) = 0$, $\frac{\partial M}{\partial v_i}\Big|_{v_1 = \cdots = v_n = t = 0} = 0, i = \overline{1, n}.$

Thus, the calculation of the function v in the representation of the desired small solution (8.1.2) to (8.1.1) is reduced to finding a solution $\bar{v} \to 0$ as $t \to 0$ from the system (8.2.4).

Remark 8.1. One can easily verify the identity $\det(-\lambda E + A) = (-1)^n L(\lambda)$, where $L(\lambda) = \lambda(\lambda-1)\ldots(\lambda-(n-1))+a_{n-1}(\epsilon)\lambda(\lambda-1)\ldots(\lambda-(n-2))+\cdots+a_0(\epsilon)$ is the characteristic polynomial of the Euler differential operator $\mathcal{L}$ in the main part of equation (8.1.3). Due to the structure of the matrix A we have

$$\operatorname{rank}(-\lambda E + A) \geq n - 1$$

with any λ.

If $\operatorname{rank}(-\lambda E + A) = n-1$ then vector $\bar{e}$, that satisfies the homogeneous system $\lambda\bar{e} = A\bar{e}$, has a Jordan chain of length p, where p is multiplicity of the root λ of characteristic polynomial $L(\lambda)$.

Theorem 8.1. *Let conditions* **A**, **B**, *and* **C** *be satisfied. Fix* $N > s\|A\|$ *and assume that among roots of characteristic polynomial* (8.1.4) *there are no numbers* $\frac{i}{s}, i = \overline{1, N}$. *Then equation* (8.1.1) *has a solution in the form*

$$x(t) = \sum_{i=0}^{N} x_i t^{\frac{r+i}{s}} + o(t^{(r+N)/s}). \tag{8.2.5}$$

Proof. Choose numbers r/s and x_0 in accordance with Conditions **A**, **B**, and **C**. We seek for a solution to the main equation (8.1.1) in the form $x = t^{r/s}(x_0 + v(t))$. Here $v(t)$ is the first component of the vector $\bar{v}$, which satisfies the system (8.2.4). We seek the vector $\bar{v}$ in the form

$$\bar{v} = \sum_{i=0}^{N} \bar{v}_i t^{i/s} + t^{N/s}\bar{\omega}(t), \tag{8.2.6}$$

where $\bar{\omega} \to 0$ as $t \to 0$.

Substituting (8.2.6) into (8.2.4) and taking into account the fact that by assumption of Theorem 8.1 $\det(nE - sA) \neq 0$ with $n \in \mathbb{N}$, we recurrently determine coefficients $\bar{v}_i$ from the linear systems

$$\left(\frac{i}{s}E - A\right)\bar{v}_i = m_i(\bar{v}_1, \ldots, \bar{v}_{i-1}), \ i = \overline{1, N}. \tag{8.2.7}$$

The right-hand sides in (8.2.7) are constructed with the method of undetermined coefficients. Since $\bar{v}_i = (v_{i1}, \ldots, v_{in})'$, in formula (8.2.5) we set $x_i = v_{i1}, i = 1, \ldots, N$. For determining the vector function $\bar{w}(t)$ we obtain the system

$$t\frac{d\bar{w}}{dt} = \left(-\frac{N}{s}E + A\right)\bar{w} + g(\bar{w}, t^{1/s}), \tag{8.2.8}$$

where $||q|| = \mathcal{O}(t^{1/s})$ with $||\bar{w}|| \leq r$, $0 \leq t \leq p$,

$$||g(\bar{w}_1, t^{1/s}) - g(\bar{w}_2, t^{1/s})|| \leq l||\bar{w}_1 - \bar{w}_2||.$$

It remains to prove that with sufficiently large N system (8.2.8) has unique solution $\bar{w} \to 0$ as $t \to 0$. One can find this solution from the equivalent integral equation

$$\bar{w}(t) = \int_0^t \tau^{-1} \exp\left[\left(-\frac{N}{s}E + A\right) \ln \frac{t}{\tau}\right] g(\bar{w}(\tau), \tau^{1/s}) \, d\tau \equiv \Phi(\bar{w}, t)$$

(8.2.9)

using the method of successive approximations with the zero initial approximation $\bar{w}_0 = 0$. Indeed, since $N > S||A||$, $\ln(\frac{t}{\tau}) \geq 0$ with $\tau \leq t$, $\text{sign}\tau = \text{sign}t$, there exists a constant value C such that with $\tau \leq t$ the following bound is valid

$$\left\|\exp\left[\left(-\frac{N}{s}E + A\right) \ln(t/\tau)\right]\right\|_{\mathcal{L}(R^n \to R^n)} \leq C \, (t/\tau)^{-\frac{N}{s} + ||A||}.$$

Therefore with $N > s||A||$ we get the following bound

$$\left\|\int_0^t \tau^{-1} \exp\left[\left(-\frac{N}{s}E + A\right) \ln(t/\tau)\right] d\tau\right\|_{\mathcal{L}(R^n \to R^n)} \leq \frac{C}{\frac{N}{S} - ||A||}.$$

Fix $q \in (0, 1)$ and choose N such that

$$\frac{cl}{\frac{N}{s} - ||A||} \leq q. \tag{8.2.10}$$

Introduce the space $\mathcal{C}_{[0,p]}$ of continuous vector functions $\bar{w}(t)$ with norm

$$||w|| = \max_{0 \leq t \leq p, \, 1 \leq i \leq n} |w_i(t)|.$$

In this space we define the set $S = \{||\bar{w}|| \leq r, |w_i(t)| \leq rt^{1/s}, i = 1, .., n\}$. Let us seek for a small solution $\bar{w}(t)$ to equation (8.2.9) in set S. With sufficiently large N in view of bound (8.2.10) we have

$$||\Phi(\bar{w}_1, t) - \Phi(\bar{w}_2, t)|| \leq q||\bar{w}_1 - \bar{w}_2||.$$

Therefore the operator Φ is contractive with $||\bar{w}|| \leq r$. Furthermore, with $||\bar{w}|| \leq r$ we have

$$||\Phi(\bar{w}, t)|| \leq ||\Phi(\bar{w}, t) - \Phi(0, t)|| + ||\Phi(0, t)||$$

$$\le qr + \max_{0 \le t \le p} \left|\left| \int_0^t \tau^{-1} \exp\left[\left(-\frac{N}{s}E + A\right)\ln\frac{t}{\tau}\right] g(\bar{w}(\tau), \tau^{1/s})d\tau\right|\right|_{\mathbb{R}^n}$$

$$\le qr + \frac{c}{\frac{N}{s} - ||A||} \max_{0 \le \tau \le p} ||q(0, \tau^{1/s})||_{\mathbb{R}^n}.$$

Since $g(0, \tau^{1/s}) \to 0$ as $\tau \to 0$, with given q and N we can choose $p > 0$ so as to fulfill the inequality $\frac{c}{\frac{N}{s}-||A||} \max_{0 \le \tau \le p} ||g(0, \tau^{1/s})||_{\mathbb{R}^N} \le (1-q)r$. Therefore the operator Φ maps the ball $||\bar{w}|| \le r$ to itself. Moreover

$$||\Phi(\bar{w}, t)|| = \mathcal{O}(t^{1/s}),$$

because

$$||\Phi(\bar{w}, t)|| \le \frac{c}{\frac{N}{s} - ||A||} ||q(\bar{w}, t^{1/s})||,$$

where

$$||g(\bar{w}, t^{1/s})|| = \mathcal{O}(t^{1/s}).$$

Therefore the contractive operator Φ maps the set $\mathcal{C}_{[0,p]}$ to itself, while system (8.2.8) has a unique solution $\bar{w} \to 0$ as $t \to 0$, provided that N is sufficiently large.   □

When solving concrete equations under assumptions of Theorem 8.1, one can seek for a solution immediately in the form of series (8.2.5); in the analytical case it converges in a neighborhood of zero. Let us weaken the conditions of Theorem 8.1, assuming that the characteristic polynomial $L(\lambda)$ has roots in the form i/s. Then the obtained result becomes more interesting. Namely, in this case equation (8.1.1) has a c-parametric family of solutions in the form (8.2.5), where the coefficients, beginning with some i, are functions of $\ln t$ depending on p arbitrary constants, p is the multiplicity of the root $\frac{i}{s}$ of the characteristic polynomial. Let us describe the way for constructing the family of small solutions in this case. Introduce the following auxiliary linear system:

$$\frac{dv}{dz} = Bv + f(z), \qquad (8.2.11)$$

where B is a constant matrix and $f(z)$ is polynomial of the order m. Consider the construction of polynomial solutions to this system. If $\det B \ne 0$, then the system (8.2.11) in the class of polynomials has the only solution $v = \sum_{i=0}^m a_i z^i$. Coefficients $a_m, a_{m-1}, \ldots, a_0$ are calculated in the indicated order by the method of undetermined coefficients. If $\det B = 0$, then for constructing a solution to system (8.2.11) in the class of polynomials it is convenient to use the Jordan normal form of the matrix B. Really, let $TBT^{-1} = J$, where $J = \{\lambda_1 E_1 + H_1, \ldots, \lambda_k E_k + H_k\}$ is the normal Jordan set, $\lambda_i E_i + H_i$ are Jordan cells of the order p_i. Some of λ_i can coincide. Let $\operatorname{rank} B = r$. Then without loss of generality we can assume that $\lambda_1 = \cdots = \lambda_{n-r} = 0$, $\lambda_i \neq 0, i = n - r + 1, \ldots, k$. In this case the following lemma is valid.

Lemma 8.1. *Let* $\operatorname{rank} B = r$ *and let a vector* $f(z)$ *be a polynomial of the order* m. *Then system* (8.2.11) *has a polynomial solution of the order* $m + \max(p_1, \ldots, p_{n-r})$, *depending on* $p_1 + \cdots + p_{n-r}$ *arbitrary constants.*

Proof. By putting $u = T^{-1}w$ in system (8.2.11) and multiplying the result by T we reduce it to the following form

$$\frac{dw}{dz} = J\omega + Tf(z). \tag{8.2.12}$$

In accordance with the structure of the matrix J we divide system (8.2.12) into k independent subsystems

$$\frac{dw_i}{dt} = (\lambda_i E_i + H_i)w_i + p_i(z), \ i = 1, \ldots, k.$$

Since $\lambda_1 = \cdots = \lambda_{n-r} = 0$, we calculate coordinates of vectors w_i, $i = 1, \ldots, n - r$ successively p_i times integrating (from zero to z) of the source part of the equation. Therefore vectors $w_i(z)$, $i = 1, \ldots, n - r$ appear to be polynomials of the order $m + p_i$ depending on p_i constants of integration. The rest of the vectors $w_i(z)$, $i = n - r + 1, \ldots, k$ are uniquely defined by means of the method of undetermined coefficients as polynomials of the mth order, because $\det (\lambda_i E_i + H_i) \neq 0$ for $i = n - r + 1, \ldots, k$. $\square$

Theorem 8.2. *Let conditions* **A**, **B**, *and* **C** *be satisfied. Assume that some of the numbers* $\frac{1}{s}, \frac{2}{s}, \cdots \frac{N}{s}$, *where* $N \leq s\|A\|$ *are roots of the characteristic polynomial* $L(\lambda)$, *denote the least of them by* l/s. *Then equation* (8.1.1) *has a solution in the form*

$$x(t) = \sum_{i=0}^{l-1} x_i t^{\frac{r+i}{s}} + \sum_{i=l}^{N} x_i(\ln t) t^{(r+i)/s} + o(t^{\frac{r+N}{s}}). \tag{8.2.13}$$

Proof. The proof of Theorem 8.2 is analogous to the proof of Theorem 8.1, but one should take into account that coefficients in solution (8.2.13) depend on $\ln t$. Therefore instead of the algebraic system (8.2.13) we obtain those

$$\left(\frac{i}{s}E - A\right)v_i = m_i(v_1, \ldots, v_{i-1}), \ i = 1, \ldots, l - 1$$

and the following systems of differential equations $(z = \ln t)$

$$\frac{dv_i}{dz} = \left(-\frac{i}{s}E + A\right)v_i + m_i(v_1, \ldots, v_{i-1}), \; i = l, l+1, \ldots, N.$$

Since

$$\det\left(\frac{i}{s}E - A\right) \neq 0$$

with $i = \overline{1, l-1}$, values $v_1, \ldots, v_{l-1}$ are determined unambiguously and are independent of z. By conditions of Theorem 8.2 we have $\det\left(-\frac{l}{s}E + A\right) = 0$. And in view of Remark 8.1

$$\text{rank}\left(-\frac{l}{s}E + A\right) = n - 1.$$

The corresponding eigenvector has a Jordan chain of length p_l, where p_l the multiplicity of the root $\frac{l}{s}$ of the characteristic polynomial $L(\lambda)$. Therefore, according to the Lemma, $v_l(z)$ is the polynomial of the order p_l depending on p_l arbitrary constants of integration. One can also construct coefficients $v_{l+1}(z), \ldots, v_N(z)$ as polynomials, whose orders are determined in accordance with the Lemma 8.1. These coefficients can contain new arbitrary constants if l/s is not the only root of the characteristic polynomial $L(\lambda)$ among numbers $(l/s, (l+1)/s, \ldots, N/s)$. $\qquad\square$

8.2.1 *Possible Generalizations*

If among roots of the characteristic polynomial there are λ, possibly, complex ones with positive real parts, which do not belong to the set $(l/s, (l+1)/s, \ldots, N/s)$, then the class of small solutions to equation (8.1.1) is wider than the set of those constructed above in the theorems; it can be extended in the class of complex-valued functions. Since coefficients of the characteristic polynomial $L(\lambda)$ are real, the polynomial $L(\lambda)$, together with a root λ, has the conjugate root $\bar{\lambda}$. Therefore partial sums of the corresponding expansions of small solutions can contain functions in the form $t^{\Re\lambda} \cos(\Im\lambda \ln t)$, $t^{\Re\lambda} \sin(\Im\lambda \ln t)$. Consider the construction of small solutions in this case. Introduce the vector $\lambda = (\lambda_1, \ldots, \lambda_l)$, where λ_i are roots of the characteristic polynomial, $\Re\lambda_i > 0$. Assume that extracted are roots λ_i, such that $\lambda_i \neq \sum\limits_{j=1, j\neq i}^{l} m_j\lambda_j + m$ with natural m, m_j. For simplicity let the function $R(x, t)$ in equation (8.1.1) be infinitely differentiable near zero and $s = 1$ in formula (8.1.2). We will seek for a solution to the reduced system (8.2.4) in the form

$$v = \sum_{j=1}^{\infty} v_{0j}(\ln t)t^j + \sum_{j=0,\,|i|\geq 1}^{\infty} v_{ij}(\ln t)t^{(\lambda,i)+j}. \tag{8.2.14}$$

Note that the second sum contains no integer powers of the argument t due to the choice of the vector λ. By means of the method of undetermined coefficients we obtain the following recurrent sequence of linear differential equations with respect to $v_{0j}(z)$, $v_{ij}(z)$, $(z = \ln t)$:

$$\frac{dv_{0j}}{dz} = (-jE + A)v_{0j} + m_{0i}(v_{01}, \ldots, v_{0j-1}), \; j = 1, 2, \ldots$$

$$\frac{dv_{ij}}{dz} = ((-(\lambda, i) + j)E + A)v_{ij} + m_{ij}(v_{rs}), \; |r| + s < |i| + j, \; i = 1, 2, \ldots,$$

$$j = 0, 1, \ldots$$

Note that $m_{i0} = 0$ with $|i| = 1$ and in view of Remark 8.1 rank $(-(\lambda, i)E + A) = n - 1$, $|i| = 1$. Therefore coefficients v_{i0} with $|i| = 1$ are determined (accurate to an arbitrary constant) as solutions to the corresponding homogeneous systems

$$(-(\lambda, i)E + A)v_{i0} = 0, |i| = 1.$$

This lemma also allows one to obtain the rest of the coefficients $v_{ij}(z)$ in expansion (8.2.14) as polynomials of z of increasing orders. As a result, we obtain the following family of small solutions to equation (8.1.1):

$$x(t) = t^r \left(x_0 + \sum_{j=0}^{\infty} v_{0j}(\ln t)t^j + \sum_{j=0}^{\infty} \sum_{|i|=1}^{\infty} v_{ij}(\ln t)t^{(\lambda,i)+j} \right). \tag{8.2.15}$$

Coefficients v_{ij} are functions of $\ln t$ and depend on l free parameters, where l is the number of roots with positive real parts of the characteristic polynomial $L(\lambda)$, taking into account their multiplicities. As in Theorem 8.1, using the principle of contracted mappings, one can prove that partial sums of the formal solution (8.2.15) are asymptotic approximations of the family of small solutions in form (8.1.2) to main equation (8.1.1). In the analytical case series (8.2.15) converges in a neighborhood of origin.

8.3 Magnetic Insulation Model Example

Consider the following differential equation:

$$\sqrt{(1 + \phi)^2 - 1} \frac{d^2\phi}{dx^2} = j(1 + \phi).$$

This model equation occurs in the analysis magnetic insulation problem of a vacuum diode. Here ϕ is the potential of the electric field and and j is the current. Here readers may refer to the paper of [Abdallah *et al.* (1997)]. Let us seek for a small continuous solution subject to $\phi(0) = 0$, $\phi'(0) = 0$. In this example the Newton diagram has one segment with vertices at points $(3/2, -2)$, $(0,0)$. Replacement (8.1.2) takes the form $\phi(x) = x^{4/3}(\phi_0 + v(x))$, where $\phi_0 = \left(\frac{9}{4\sqrt{2}}j\right)^{3/2}$, while $v(x)$ satisfies the equation

$$x^2 \frac{d^2v}{dx^2} + \frac{8}{3}x\frac{dv}{dx} + \frac{2}{3}v = M(v, x^{1/3})$$

with the Euler differential operator of the 2nd order in the main part, $M(0,0) = 0$, $M_v'(0,0) = 0$. The corresponding characteristic polynomial $\lambda^2 + \frac{5}{3}\lambda + \frac{2}{3} = 0$ has two negative roots $\lambda_1 = -2/3$, and $\lambda_2 = -1$. Therefore on the base of Theorems 8.1 and 8.2 in a neighborhood of the point $x = 0$ exactly one small real solution $\phi(x) = \left(\frac{9}{4\sqrt{2}}j\right)^{2/3} x^{4/3} + \mathcal{O}(x^{5/3})$ satisfies our equation with arbitrary current j.

Chapter 9

Convex Majorants Method in the Theory of Nonlinear Volterra Equations

9.1 Problem Statement. The State of the Art

Let us consider the following nonlinear continuous operator

$$\Phi(\omega_1, \ldots, \omega_n, u, t) : E_1 \times \cdots \times E_1 \times \mathbb{R}^1 \to E_2$$

of $n+1$ variables $\omega_1, \ldots, \omega_n, u$, which are abstract continuous functions of real variable t with values in E_1. Here E_1, E_2 are Banach spaces and

$$\Phi(0, \ldots, 0, u_0, 0) = 0, \; K_i : \underbrace{\mathbb{R}^1 \times \ldots \mathbb{R}^1}_{i+1} \times \underbrace{E_1 \times \cdots \times E_1}_{i} \to E_2$$

are nonlinear continuous operators depending on $u(s) = (u(s_1), \ldots, u(s_n))$ and $t, s_1, \ldots s_n$ are real variables. Let

$$\omega_i(t) = \int_0^t \cdots \int_0^t K_i(t, s_1, \ldots, s_i, u(s_1), \ldots, u(s_i)) \, ds_1 \ldots ds_i, \; i = \overline{1, n}$$

and let us address the following operator-integral equation for $t \in [0, T)$

$$F(u, t) \equiv \Phi\left(\int_0^t K_1(t, s, u(s)) ds, \int_0^t \int_0^t K_2(t, s, s_1, s_2, u(s_1), u(s_2)) ds_1 ds_2, \right.$$

$$\left. \cdots \int_0^t \cdots \int_0^t K_n(t, s_1, \ldots s_n, u(s_1), \ldots u(s_n)) \, ds_1 \ldots ds_n, u(t), t \right)$$

$$= 0. \tag{9.1.1}$$

Unknown abstract continuous function $u(t)$ maps into E_1. Our objective is to find the continuous solution $u(t) \to u_0$ as $t \to 0$. For $E_1 = E_2 = \mathbb{R}^1$ the equation (9.1.1) appears in *feed-back* control flow process based on the Volterra models it is necessary to find input signal $x(t)$ based on known $f(t)$, i.e., to solve the nonlinear VIE of the 1st kind which is the special kind of the equations addressed in present chapter (here readers may also refer to the articles of [Apartsyn (2013); Belbas and Bulka (2010)]). To the best of our knowledge, the equation (9.1.1) has not yet been studied in general case of the Banach spaces E_1, E_2.

One of the common constructive methods in the theoretical and applied studies is method of majorants. Leonid V. Kantorovich (here readers may refer to the works of [Kantorovich (1937); Kantorovich *et al.* (1950)]) studied the functional equations in B_K–spaces and converted the classical method of majorants into its abstract form, which makes it's methodology more clear and more unified. In his monograph he specifically outlined the role of the *main* solutions of nonlinear equations and role of the corresponding majorants. The main solutions are unique by definition and can be constructed using the successive approximations from the equivalent equation (9.2.1) starting from zero initial estimate. The main continuous solution $u^+(t)$ in the points of the interval $[0, T)$ satisfies the bound $||u^+(t)||_{E_1} \le z(t)$ where $z(t)$ is continuous positive solution of the following majorant VIE

$$z(t) = f\left(\int_0^t \gamma(z(s))ds\right), \quad z(t) \in C_{[0,T)}^+. \tag{9.1.2}$$

Here and below f, γ are monotone increasing continuous functions. If the trivial solution $u = 0$ satisfies the equation (9.1.1) then this solution is the main solution. This trivial case is excluded from the consideration below. If one continue the main nontrivial solution $u^+(t)$ outside of the interval $[0, T)$ (where we observe the convergence of the successive approximations) in the right hand side direction from the margin point T then the solution $u^+(t)$ can go to ∞ or start branching. Here readers may refer to book of [Zeidler (1986)]. Obviously, there is a case when operator F satisfies the Lipschitz condition for any u and the main solution is continuable on the whole interval $[0, \infty)$. If the Lipschitz condition is not satisfied, then in addition to the main solution the equation (9.1.1) can have arbitrarily many other continuous solutions which cross the main solution.

Example 9.1.

$$u(t) = p\int_0^t u^{\frac{p-1}{p}}(s), \quad 1 < p < \infty.$$

Here $u_1(t) = 0$ is the main solution. Other continuous solutions: $u_2(t) = t^p$,

$$u_c(t) = \begin{cases} 0, & -\infty \le t \le c \\ (t-c)^p, & c \le t < \infty. \end{cases}$$

The main solution $u^+(t) = 0$ of this example is singular solution for the Cauchy problem $\dot{u} = pu^{\frac{p-1}{p}}$, $u(0) = 0$.

The objective of this chapter is to construct main solutions for the equation (9.1.1) on the maximal interval $[0, T)$.

Outline

The chapter consists of two sections including the illustrative examples. In Section 9.2 for equation (9.1.1) existence theorem is derived for the main solution $u(t) \to u_0$ as $t \to 0$ with estimate $||u(t)||_{E_1}$, $t \in [0, T)$.

We propose the approach for construction of the approximations $u_n(t)$ and the interval $[0, T)$ on which they converge point-wise for $\forall u_0(t)$, if $||u_0(t)||_E \leq z^+(t)$, where $z^+(t)$ is the main nonnegative solution of the corresponding majorant integral equation (9.1.2). The sufficient conditions are derived if $\lim_{t \to T} z^+(t) = \infty$ (or $\lim_{t \to T} \frac{dz^+(t)}{dt} = +\infty$), i.e., the main solution of the majorant equation (or its derivative) has the *blow-up limit* (goes to ∞ for finite time T). Under such a conditions the unknown solution $u(t)$ of the equation (9.1.1) can also go to infinity during the finite time $T' \geq T$ or appear to be branching .

In the Section 9.3 of this chapter we demonstrate how to construct and employ the following majorant algebraic systems

$$\begin{cases} r = R(r, t), \\ 1 = R'_r(r, t) \end{cases} \tag{9.1.3}$$

for construction of the main solution of the equation (9.1.1). In the algebraic system (9.1.3) $R(0, 0) = 0$, $R'_r(0, 0) = 0$, $R(r, t)$ is the convex function w.r.t. r. The algebraic majorant systems (9.1.3) by [Grebennikov and Ryabov (1979)] were also called as the Lyapunov majorants. Such majorants as well as more general algebraic majorants were used in mechanics (readers may refer to the book of [Sidorov *et al.* (2002)]) and for the construction of implicit functions in spaces B_K. It is to be noted that algebraic majorant systems has the unique positive solution r^*, T^*.

Using this approach one can define the *guaranteed interval* $[0, T^*]$, on which equation (9.1.1) possesses the main solution $u(t) \to 0$ as $t \to 0$ and radius of the sphere $S(0, r^*)$ in the space $C_{[0,T^*]}^{E_1}$, in which the main solution can be constructed using the successive approximations which converge uniformly.

9.2   Integral Majorants Method

It this section we employ integral majorants for construction of the main solution. Let us consider the following equation

$$u = \mathcal{L}(u), \qquad (9.2.1)$$

where $\mathcal{L}(u) = A^{-1}(Au - F(u,t))$, which is equivalent to equation (9.1.1). Here A is continuously invertible operator from E_1 into E_2. If the operator F has the Fréchet derivative $F_u(0,0)$ and its invertible then we can assume $A = F_u(0,0)$.

Definition 9.1. If the approximations $u_n(t) = \mathcal{L}(u_{n-1})$, $u_0 = 0$ for $t \in [0,T^*)$ tends to the solution $u^+(t)$ of equation (9.2.1), then function $u^+(t)$ we call *Kantorovich main solution* of the equation (9.1.1).

It is to be noted that here we follow the monograph of Leonid V. Kantorovich, where the term *"the main solution of functional equation"* has been formulated, here readers may refer to the books of [Kantorovich *et al.* (1950); Kantorovich (1937)]. Under the solution we will be assuming the main solution below.

Let us study the operator $F(u,t) - Au : \mathcal{C}^{E_1}_{[0,T]} \to \mathcal{C}^{E_2}_{[0,T]}$. Here $\mathcal{C}^{E_1}_{[0,T]}$ and $\mathcal{C}^{E_2}_{[0,T]}$ are complete spaces. Then we get the following bounds using norms of the spaces E_1 and E_2 as follows:

A. $\|F(u,t) - Au\|_{E_2} \le f(\int_0^t \gamma(\|u(s)\|_{E_1}) ds)$, $t \in [0,T)$.

Let in the inequality **A** and below the following assumption be hold:

B. The functions γ, f are continuous and monotone increasing on the segments $[0,z']$ and $[\gamma(0), \gamma(z')]$, $z \le \infty$ correspondingly;

C. For $t \in [0,T)$ exist the function $z'(t)$ in the cone $\mathcal{C}^+_{[0,T]}$ such as

$$z'(t) \ge f\left(\int_0^t \gamma(z'(s)) ds \right). \qquad (9.2.2)$$

Remark 9.1. In the Lemmas 9.2 and 9.3 we propose the method to define the margin T such as for $t \in [0,T)$ the condition **C** will be satisfied. Because of the condition **A** $f(\gamma(0)t) \ge 0$. Zero is lower solution of the majorant integral equation (9.1.2), and $z'(t)$ is upper solution in the cone $\mathcal{C}^+_{[0,T)}$.

Under the conditions **C** and **B** we introduce the sequence

$$z_n(t) = f\left(\int_0^t \gamma(z_{n-1}(s))ds\right), \; z_0 = 0.$$

Then due to the Theorem 2.11 (from the book of [Kantorovich *et al.* (1950)], p.464) for any n, $t \in [0, T)$ the inequalities

$$0 = z_0(t) \le z_1(t) \le \cdots \le z_n(t) \le z'(t)$$

are satisfied. Hence the limit $\lim_{n\to\infty} z_n(t) = z^+(t)$ is attained. Since γ, f are continuous functions and due to the Lebesgue theorem (readers may refer here to the book of [Shilov and Gurevich (1978)]) the limit is attained

$$\lim_{n\to\infty} f\left(\int_0^t \gamma(z_n(s))ds\right) = f\left(\int_0^t \gamma(z^+(s))ds\right).$$

Thus function $z^+(t)$ appears to be continuous on $[0, T)$ and to be the main solution of the majorant equation (9.1.2). Approximation $z_n(t)$ in the points of the interval $[0, T)$ converge to $z^+(t)$, $z^+(t) \in \mathcal{C}_{[0,T]}^+$.

Let us now proceed to the construction of the solution $u^+(t)$ of the equation (9.1.1) using the successive approximations. In addition to **A, B** and **C** let the following inequality be satisfied

D. $||F(u+\Delta u, t) - F(u, t) - Au||_{E_2} \le f\left(\int_0^t \gamma(||u(s)||_{E_1} + ||\Delta u(s)||_{E_1})ds\right) - f(\int_0^t \gamma(||u(s)||_{E_1})ds).$

Under the condition of the Fréchet differentiability of the operators $F(u, t)$, $f(\int_0^t \gamma(z(s))ds)$ the verification of inequality **D** can be replaced with verification of the condition **E** (see below).

Indeed, let exist the continuous Fréchet derivatives of these operators for $t \in [0, T)$ in the norms of linear bounded operators in the spaces $\mathcal{L}(E_1 \to E_2)$ and $\mathcal{L}(\mathcal{C}_{[0,T)}^+ \to \mathcal{C}_{[0,T)}^+)$ correspondingly. Under such assumption we assume functions f, γ enjoy monotone increasing and continuous derivatives and the Fréchet differential f is defined by the formula

$$f_z'\left(\int_0^t \gamma(z(s))ds\right)h \equiv f_\gamma'\left(\int_0^t \gamma(z(s))ds\right)\int_0^t \gamma_z'(z(s))h(s)ds$$

for any $h(s) \in \mathcal{C}_{[0,T]}^+$.

Let in addition to conditions **A** and **B** $\forall V(t) \in \mathcal{C}_{[o,T]}^{E_1}$ the following inequality be satisfied

E. $||(F_u(u,t) - A)V||_{E_2} \leq f_z'\left(\int_0^t \gamma(||u(s)||_{E_1} ds\right)||V||_{E_1}$, $||V||_{E_1} \in \mathcal{C}_{[0,T)}^+$.

Then we have the following Lemma.

Lemma 9.1. *Let the inequality* **E** *be satisfied and the derivatives* f_γ', γ_z' *are monotone increasing. Then inequality* **D** *will be satisfied.*

Proof. Let us employ the Lagrange finite-increments formula ([Trénoguine (1985)], p.367) and conditions of the Lemma 9.1. Then we get the inequality

$$||F(u + \Delta u, t) - F(u,t) - A\Delta u||_{E_2}$$

$$= ||\int_0^1 (F_u(u + \Theta\Delta u, t) - A)d\Theta\Delta u||_{E_2} \leq \int_0^1 f_\gamma'\left(\int_0^t \gamma(||u(s)||_{E_1}\right.$$

$$\left. + \Theta||\Delta u(s)||_{E_1})ds\right) \int_0^t \gamma'\left(||u(s)||_{E_1} + \Theta||\Delta u(s)||_{E_1}\right)||\Delta u(s)||_{E_1}dsd\Theta$$

$$= f\left(\int_0^t \gamma(||u(s)||_{E_1} + ||\Delta u(s)||_{E_1})ds\right) - f\left(\int_0^t \gamma(||u(s)||_{E_1})ds\right).$$

$\square$

Let us now construct the approximations $u_n(t) = \mathcal{L}(u_{n-1})$, $u_0 = 0$ to solution $u^+(t)$. We follow the proof of Theorem 2.22 ([Kantorovich *et al.* (1950)], p.466) and state the bounds $||u_{n+p}(t) - u_n(t)||_{E_1} \leq z_{n+p}(t) - z_n(t)$ for $t \in [0,T)$, where $z_n(t) = f(\int_0^t \gamma(z_{n-1}(s))ds)$, $z_n(t) \in \mathcal{C}_{[0,T)}^+$, $u_n(t) \in \mathcal{C}_{[0,T)}^{E_1}$, $u_0 = 0, z_0 = 0$.

It is to be noted here that similar bounds in different problem has been also used in the monograph of [Sidorov *et al.* (2002)] for studies of explicit mappings based on the convex majorants method.

Due to the conditions **A**, **B** and **C** and based on on the above mentioned approach the limit $\lim_{n\to\infty} z_n(t) = z^+(t)$ exists for $\forall t \in [0,T)$, i.e., $z_n(t)$ is fundamental sequence in the each point $t \in [0,T)$. Hence the sequence of abstract functions $u_n(t)$ with values in the Banach space E_1 for each $t \in [0,T)$ converges in norms of the space E_1 to function $u^+(t)$. Since the operator $\mathcal{L}(u)$ is continuos, the equality $u^+(t) = \mathcal{L}(u^+)$ is satisfied, i.e., $u^+(t)$ satisfies the equation (9.1.1) and belongs to the space $\mathcal{C}_{[0,T)}^{E_1}$.

Hence we have the following theorem.

Theorem 9.1. *Let for* $t \in [0, T)$ *conditions* **A, B, C** *and* **D** *are satisfied. Then equation* (9.1.1) *in the space* $C^{E_1}_{[0,T)}$ *has the main solution* $u^+(t)$. *Moreover,* $\|u^+(t)\|_{E_1} \leq z^+(t)$, *were* $z^+(t)$ *is main solution of majorant equation* (9.1.2), *approximations* $u_n(t) = \mathcal{L}(u_{n-1})$, $u_0 = 0$ *converge to* $u^+(t)$ *in norm of the space* E_1 *for any* $t \in [0, T^+)$, *approximations* $z_n(t) = f(\int_0^t \gamma(z_{n-1}(s))ds)$, $z_0 = 0$ *converge to* $z^+(t)$.

In Theorem 9.1 T^+ remains not defined. For the definition of T^+ we reduce the majorant integral equation (9.1.2) to the Cauchy theorem for separable differential equation. For this objective we introduce the differentiable function $w(t) = \int_0^t \gamma(z(s))ds$. Then $\frac{dw(t)}{dt} = \gamma(z(t))$, $w(0) = 0$, where $z(t) = f(w(t))$. That is why the Cauchy problem which is equivalent to the equation (9.1.2) is following

$$\begin{cases} \dfrac{dw}{dt} = \gamma(f(w(t))), \\ w(0) = 0. \end{cases} \tag{9.2.3}$$

Lemmas 9.2 and 9.3 define the estimate of the interval $[0, T^+)$, on which the Cauchy problem (9.2.3) possesses the unique solution $w^+(t)$ in space $C^+_{[0,T^+)}$ and approximations $w_n(t) = \int_0^t \gamma(f(w_{n-1}(s)))ds$, $w_0 = 0$ converge to this unique solution.

Lemma 9.2. *Let* $\gamma(f(w))$ *be continuos, strictly positive and monotone increasing function. Let exists* $\lim_{w \to \infty} \int_0^w \frac{dw}{\gamma(f(w))} = T^+$. *Then* (9.2.3) *in cone* $C^+_{[0,T^+]}$ *has monotone increasing solution* $w^+(t)$. *The approximations* $w_n(t) = \int_0^t \gamma(f(w_{n-1}(s)))ds$, $w_0 = 0$ *converge to* $w^+(t)$, $\lim_{t \to T^+} w^+(t) = \infty$.

Proof. Let us separate the variables in (9.2.3) and reduce the Cauchy problem to search for the positive monotone increasing branch of implicit function $w = w(t)$, $w(0) = 0$ from equation $\Phi(w) = t$, where $\Phi(w) = \int_0^w \frac{dw}{\gamma(f(w))}$. If $\gamma(f(w))$ is rational fraction then $\Phi(w)$ can be explicitly constructed in terms of logarithms, arctangent and rational functions. It is to be noted that under conditions of the Lemma 9.1, function $\Phi(w)$ is continuous and monotone increasing on semi-axis $[0, \infty)$, i.e., $\Phi' = \frac{1}{\gamma(f(w))} > 0$, $\lim_{w \to 0} \Phi(w) = 0$, $\lim_{w \to \infty} \Phi(w) = T^+$. Hence the mapping $\Phi : [0, \infty) \to [0, T^+)$

is bijective, equation $\Phi(\omega) = t$ for $0 \leq t < T^+$ uniquely defines function $\omega^+(t)$, which obviously satisfies the integral equation

$$\omega(t) = \int_0^t \gamma(f(\omega(s)))ds.$$

Because of the monotone increasing of the functions f and γ, the approximations $\omega_n(t) = \int_0^t \gamma(f(\omega_{n-1}(s)))ds$, $\omega_0 = 0$ for $t \in [0, T^+)$ converge to $\omega^+(t)$. $\square$

If $\gamma(f(\omega))$ is rational fraction, then in number of cases the solution $\omega^+(t)$ can be explicitly constructed using the computer algebra systems, here readers may refer to the article of [Apartsyn (2013)].

Remark 9.2. For known $\omega^+(t)$ using the formula $z^+(t) = f(\omega^+(t))$ we find the solution of majorant integral equation (9.1.2). It is to be noted that under the conditions of the Lemma 9.2, approximations $z_n = f(\int_0^t \gamma(z_{n-1}(s)ds))$, $z_0 = 0$ converge to the solution $z^+(t)$, $t \in [0, T^+)$.

Remark 9.3. If under the conditions of Lemma 9.2 $\lim\limits_{\omega \to \infty} \int_0^\omega \frac{d\omega}{\gamma(f(\omega))} = \infty$ then the solution $z^+(t)$ is continuable on $[0, \infty)$. This result follows from the Theorem 2.7 (here readers may refer to the book of [Barbashin (1970)], p.148).

For example, let inequality

$$||F(u, t) - Au||_{E_2} \leq a \int_0^t ||u(s)||ds + b, \ a > 0, \ b > 0,$$

be satisfied for $\forall u$, $0 \leq t < \infty$. Then majorant integral equation (9.1.2) will be linear as follows $z(t) = a \int_0^t z(s)ds + b$ and possesses the unique solution $z(t) = be^{at}$, $0 \leq t < \infty$. It is to be noted that is this case $\gamma(f(\omega)) = a\omega + b$, $\lim\limits_{\omega \to \infty} \int_0^\omega \frac{d\omega}{a\omega + b} = \infty$. If in this case

$$||F(u + \Delta u, t) - F(u, t) - A\Delta u||_{E_2} \leq a \int_a^t ||\Delta u(s)||_{E_1}ds,$$

then conditions of the theorem are satisfied on semi-axis $0 \leq t < \infty$ and equation (9.1.1) will possesses the solution $u^+(t)$ in the space $C_{[0, \infty)}^{E_1}$,

$||u^+(t)||_{E_1} \leq be^{at}$. Obviously, for $||u(t)||_{E_1} \geq be^{at}$ the equation (9.1.1) may possess another solutions.

Lemma 9.3. *Let superposition $\gamma(f(\omega))$ be continuous and strictly positive for $0 \leq \omega \leq \omega^*$. Let the following limits $\lim\limits_{\omega \to \omega^*} \gamma(f(\omega)) = \infty$, $\lim\limits_{\omega \to \omega^*} \int_0^\omega \frac{d\omega}{\gamma(f(\omega))} = T^+$ are exist. Then Cauchy problem (9.2.3) for $t \in [0, T^+]$ in the cone $C_{[0,T^+]}^+$ has continuous monotone increasing solution $\omega^+(t)$ and $\lim\limits_{t \to T^+} \frac{d\omega^+}{dt} = 0$. Moreover, an approximations*

$$\omega_n(t) = \int_0^t \gamma(f\omega_{n-1}(s))ds$$

converge for $0 \leq t \leq T^+$, $\omega_0 = 0$ to the solution $\omega^+(t)$.

Proof of the Lemma 9.3 follows from the bijective mapping $\Phi : [0, \omega^*] \to [0, \Phi(\omega^*)]$ for $\Phi(\omega) = \int_0^\omega \frac{d\omega}{\gamma(f(\omega))}$, $\Phi(\omega^*) = T^+$.

Remark 9.4. Under the conditions of Lemma 9.3 the point T^+ is blow-up limit of the derivative of solution $z^+(t)$ of majorant equation (9.1.2).

9.3   Algebraic Majorants Method

It this section we employ the algebraic majorants for construction of the main solution. Let in equation (9.1.1) $u_0 = 0$, i.e., $\Phi(0,\dots,0) = 0$. Our objective is to construct continuous solution $u^+(t)$ with successive approximations $u_n(t) = \mathcal{L}(u_{n-1})$ in close interval $[0, T^+]$. In the space $C_{[0,T^+]}^{E_1}$ we introduce the norm $||u|| = \max\limits_{0 \leq t \leq T^+} ||u(t)||_{E_1}$. We suppose the operator F be Fréchet differentiable (w.r.t. u). Let for $0 \leq t \leq T$, $T \geq T^+$ for $u \in S(0, r) \subset E_1$, the following inequalities be satisfied:

A'. $||F(u,t) - Au||_{E_2} \leq f(r,t);$

E'. $||F'_u(u,t) - A||_{E_2} \leq f'_r(r,t);$

G. Let functions $f(r,t)$, $f'_r(r,t)$ be positive for $r > 0$, $t > 0$ and be monotone increasing functions, $f(0,0) = 0$, $f'_r(0,0) \in [0,1)$, and let function $f(r,t)$ be convex w.r.t. r.

Then algebraic equation $r = ||A^{-1}||f(r,t)$ is the Lyapunov majorant for operator $\mathcal{L}(u)$. Since functions $f(r,t)$, $f_r(r,t)$ are monotonically increasing function and since the function $f(r,t)$ is convex, the system

$$\begin{cases} r = ||A^{-1}||f(r,t), \\ 1 = ||A^{-1}||f'_r(r,t) \end{cases}$$

has unique positive solution r^+, T^+. Moreover, equation $r = ||A^{-1}||f(r,t)$ where $0 \le t \le T^+$, possesses uniquely defined monotone increasing solution $r^* = r(t)$. Approximations $r_n(t) = ||A^{-1}||f(r_{n-1}(t),t)$, $r_0 = 0$, when $0 \le t \le T^+$ converge to the function $r(t)$. Corresponding approximations $r_n = ||A^{-1}||f(r_{n-1},T^+)$, $r_0 = 0$, converge to r^+. Function $r(t)$ is main solution of the Lyapunov majorant equation. On the base of Lemma 5.1 from the monograph of [Barbashin (1970)], p.206 if $||u_i(t)||_{E_1} \le r_i$, $i = 1,2$, $||u_2(t) - u_1(t)|| \le r_2 - r_1$, then for $0 \le t \le T^+$

$$||\mathcal{L}(u_2) - \mathcal{L}(u_1)||_{E_1} \le ||A^{-1}||(f(r_2,t) - f(r_1,t)).$$

Apart from approximations $r_n(t)$ for solutions for Lyapunov majorante, we introduce approximation $u_n(t) = \mathcal{L}(u_{n-1})$, $u_0 = 0$ of the main solution of the equation (9.1.1). For arbitrary k and $l \ge k$ because of the conditions $\mathbf{A'}$, $\mathbf{E'}$ and above mentioned inequality, we come to the bound $||u_l(t) - u_k(t)||_{E_1} \le r_l(t) - r_k(t) \le r_l(T^+) - r_k(T^+)$. Such that $r_l(T^+)$ is monotone increasing sequence and $\lim_{l\to\infty} r_l(T^+) = r^+$, then $||u_l(t) - u_k(t)||_{E_1} \le \varepsilon$ for $l,k \ge N(\varepsilon)$ if $t \in [0,T^+]$. Hence $||u_l(t) - u_k(t)||_{E_1} \le \varepsilon$ for $l,k \ge N(\varepsilon)$. Because of complete space $\mathcal{C}^{E_1}_{[0,T^+]}$ exist limit $\lim_{l\to\infty} u_l(t) = u^+(t)$. Moreover, $u^+(t)$ and approximation $u_n(t) = \mathcal{L}(u_{n-1})$, $u_0 = 0$ converge on segment $[0,T^+]$ uniformly (w.r.t. t).

Then follows

Theorem 9.2. *Let* $\Phi(0,\ldots,0) = 0$, *inequalities* $\mathbf{A'}$, $\mathbf{E'}$ *are satisfied when* $t \in [0,T^+]$, *pair* (r^+,T^+), $r^+ > 0$, $T^+ > 0$ *satisfies algebraic system*

$$\begin{cases} r = ||A^{-1}||f(r,t), \\ 1 = ||A^{-1}||f'_r(r,t), \end{cases}$$

where function $f(r,t)$ *satisfies the condition* $\mathbf{G}$. *Then equation (9.1.1) possesses the continuous solution* $u^+(t)$ *in space* $\mathcal{C}^{E_1}_{[0,T^+]}$. *Moreover, approximations* $u_n(t) = \mathcal{L}(u_{n-1})$ *converge uniformly (w.r.t. t) and* $\max_{0\le t\le T^+} ||u^+(t)|| \le r^+$.

Example 9.2. Let us consider the following problem

$$
\begin{cases}
\dfrac{\partial^2 u(x,t)}{\partial x^2} + \displaystyle\int_0^t \sin(t - \tau + x) u^2(x,\tau)d\tau = t, \\
u\big|_{x=0} = u\big|_{x=1} = 0, \ \ 0 \le x \le 1, \, t \ge 0.
\end{cases}
$$

We search for classical solution $u \to 0$ as $t \to 0$. Here $E_1 = \overset{\circ}{C}\,^{(2)}_{[0,1]}$ is space of twice differentiable w.r.t. x functions are zero on the margins $[0,1]$, $E_2 = C_{[0,1]}$. $Au = \frac{\partial^2 u}{\partial x^2}$, operator $A \in \mathcal{L}(E_1 \to E_2)$ has bounded reverse operator $A^{-1} = \int_0^1 G(x,s)[\cdot]ds$, where

$$
G(x,s) = \begin{cases}
x(s-1),\, 0 \le x \le s \le 1, \\
s(x-1),\, s \le x \le 1,
\end{cases}
$$

$$||A^{-1}||_{\mathcal{L}(E_1 \to E_2)} \le 1.$$

Following the Theorem 9.1, the corresponding majorant integral equation (9.1.2) is following $z(t) = \int_0^t z^2(s)ds + t$. Then function $z^+(t) = \tan t$ for $0 \le t < \frac{\pi}{2}$ is the main solution of majorant integral equation. Therefore $\frac{\pi}{2}$ is point, in which solution has blow-up limit $z^+(t)$. Boundary problem (due to Theorem 9.1) has the solution $u^+(x,t)$ in space $C^{E_1}_{[0,\frac{\pi}{2})}$ and $0 \le t < \frac{\pi}{2}$

$$
\max_{0 \le x \le 1} \left(\left| \frac{\partial^i u^+(x,t)}{\partial x^i} \right|, i = 0,1,2 \right) \le \tan t.
$$

From other hand, if we follow Theorem 9.1, we construct majorant algebraic equation $r = tr^2 + t$. Function $r^+(t) = \frac{1 - \sqrt{1-4t^2}}{2t}$ for $t \in [0,\, 0.5]$ is the main solution of majorant algebraic equation.

According to Theorem 9.1 we construct the following system

$$
\begin{cases}
r = tr^2 + t \\
1 = 2tr
\end{cases}
$$

which has one positive solution $T^+ = 0.5$, $r^+ = 1$. Therefore based on Theorem 9.2 we get guaranteed interval (w.r.t. t) of existence of the solution $u^+(x,t)$ with following estimate

$$
\max_{x \in [0,1],\, t \in [0,0.5]} \left\{ \left| \frac{\partial^i u^+(x,t)}{\partial x^i} \right|, i = 0,1,2 \right\} \le 1, \ ||u^+||_{E_1} \le r^+(t), \, 0 \le t \le 0.5.
$$

Since $0.5 < \frac{\pi}{2}$ then in this example the integral majorant provides more precise estimate u^+, comparing to the algebraic one.

As the footnote let us notice that with studies of the equation (9.1.1) in B_K spaces and with introduction of the abstract norms in the

Kantorovich sense it is possible to get more fine systems of majorant integral and algebraic equations. Such a majorants will characterize the solution of equation (9.1.1) more deeply. Majorant algebraic equations possible to construct and it is possible to study the solutions of the n-dimensional Volterra equations (9.1.1), namely for $t \in \mathbb{R}^n$, $n \geq 2$. As a matter of fact, the algebraic majorants provides more rough estimates comparing to the integral majorants, but it is easier to construct and to employ the algebraic majorants. Since the solution of the majorant integral equation has blow-up limit, for numerical solution in the neighborhood of such points it make sense to employ the adaptive meshes.

9.4   Application of the Convex Majorants Method

Let us demonstrate the efficiency of the proposed convex majorants method. We apply the convex majorants method to study the following nonlinear Volterra integral equation of the second kind

$$x(t) = \int_0^t K(t, s, x(s))\, ds,\ 0 < t < T < \infty,\ x(0) = 0. \tag{9.4.1}$$

Definition 9.2. [Kantorovich *et al.* (1950)] Continuous function $x(t)$, satisfying the equation (9.4.1), we call as *Kantorovich main* solution if the sequence

$$x_n = \int_0^t K(t, s, x_{n-1}(s))\, ds,\ x_0(t) = 0$$

converges to function $x(t)\ \forall t \in (0, T)$. If in addition $\lim\limits_{t \to T} |x(t)| = +\infty$, then solution *blows up* at the point T.

Let us find the guaranteed interval $[0, T)$ where exists the main solution such as the blow-up may occur if one continue solution onto $[T, +\infty)$. Beside, one must find the positive continuous function $\hat{x}(t)$, defined on $[0, T)$ such as for the main solution $x(t)$ the following a priory estimate $|x(t)| \leq \hat{x}(t)$ if satisfied for $t \in [0, T)$.

We employ the classical approach by L.V. Kantorovich (here readers may refer to Chapter 12 of the book of [Kantorovich *et al.* (1950)]). For the majorizing equations construction we will use the algorithm proposed in this chapter. Let us introduce the following conditions

A. Let function $K(t, s, x)$ be defined, continuous and differentiable wrt x in $D = \{0 < s < t < T, T \le \infty, |x| < \infty\}$.

B. Let us assume we can construct functions $m(s)$, $\gamma(x)$ which are continuous, positive and monotonically increasing functions defined for $0 < s < \infty$, $0 < x < \infty$, such as in D for any t from $[0, \infty)$ the following inequalities are satisfied

$$|K(t, s, x)| \le m(s)\gamma(|x|),$$

$$|K'(t, s, x)| \le m(s)\gamma'(|x|).$$

Here we exclude the case of $\gamma(0) = 0$ since in that case equation (9.4.1) has only trivial solution. Such solution is the main according to the Kantorovich definition. Below the functions $m(s)$, $\gamma(x)$ are assumed positive, monotone increasing, and $\gamma(x)$ is assumed convex w.r.t. x.

Integral Majorizing Equation

Let us introduce majorizing integral equation

$$\hat{x}(t) = \int_0^t m(s)\gamma(\hat{x}(s))\, ds, \tag{9.4.2}$$

which is equivalent to the Cauchy problem for the differential equation with separable variables:

$$\begin{cases} \dfrac{d\hat{x}}{dt} = m(t)\gamma(\hat{x}(t)) \\ \hat{x}|_{t=0} = 0. \end{cases} \tag{9.4.3}$$

Thus, the solution $\hat{x}(t)$ of integral equation (9.4.2) satisfies the equation $\Phi(x) = M(t)$, where $\Phi(x) = \int_0^x \frac{dx}{\gamma(x)}$, $M(t) = \int_0^t m(t)\, dt$. Because of monotone increasing positive continuous function $\Phi(x)$ there exists inverse mapping Φ^{-1} with the domain $[0, \infty)$, if $\lim_{x \to \infty} \int_0^x \frac{dx}{\gamma(x)} = +\infty$, and with the domain $[0, l)$, if $\lim_{x \to \infty} \int_0^x \frac{dx}{\gamma(x)} = l$. Thus in the first case the Cauchy problem has unique positive solution $\hat{x}(t)$ in $C_{[0,\infty)}^{(1)}$, and in $C_{[0,l)}^{(1)}$ in the second case.

Solution $\hat{x}(t)$ can be constructed with successive approximations as solution of $\Phi(x) - M(t) = 0$. Indeed, equation $\Phi(t) - M(t) = 0$ defines $\hat{x}(t)$ as explicit continuous function $\hat{x}(t) \to 0$ for $t \to 0$, since

$$\frac{d}{dx}(\Phi(x) - M(t)) = \frac{1}{\gamma(x)} \ne 0.$$

Hence the solution $\hat{x}(t)$ can be constructed with successive approximations:

$$x_n(t) = x_{n-1}(t) - \gamma(0)\left[\Phi(x_{n-1}(t)) - M(t)\right], \ x_0 = 0$$

on the small interval $[0, \Delta]$, $\Delta > 0$.

Constructed solution can be continued on the whole domain of Φ^{-1} by repeated application of the implicit theorem application.

Let $\gamma(x)$ be polynomial with positive coefficients and $\gamma(0) \neq 0$. In such case function $\Phi(x)$ can be explicitly constructed in terms of logarithms, arc-tangents and rational functions, which allows us in basic cases to construct Φ^{-1} and to explicitly build $\hat{x}(t)$.

In general case in order to build $\hat{x}(t) = \Phi^{-1}(M(t))$, satisfying equation (9.4.2), one may employ the following Lemma.

Lemma 9.4. *If* $\lim\limits_{x \to +\infty} \int_0^x \frac{dx}{\gamma(x)} = +\infty$*, then majorizing equation (9.4.2) has for* $t \in [0, \infty)$ *continuous solution* $\hat{x}(t)$*. Moreover, the sequence*

$$x_n(t) = \int_0^t m(s)\gamma(x_{n-1}(s))\,ds, \ x_0(t) = 0 \qquad (9.4.4)$$

converges for $t \in [0, \infty)$ *to function* $\hat{x}(t)$*.*

Proof. Existence of the solution $\hat{x}(t)$ of equation (9.4.2) follows from above proved solution existence of equivalent Cauchy problem (9.4.3). Herewith the sequence $\{x_n(t)\}$ will be monotone increasing and bounded above since $\hat{x}(t)$ satisfies the equation (9.4.1). Hence $\{x_n(t)\}$ has limit. Thus, the sequence $\{x_n(t)\}$ is fundamental in the space $\mathcal{C}_{[0,T_1]}$ for $T_1 < \infty$. The space $\mathcal{C}_{[0,T_1]}$ is complete and convergence is uniform for $T_1 < \infty$.

$\square$

Lemma 9.5. *Let* $\lim\limits_{x \to +\infty} \int_0^x \frac{dx}{\gamma(x)} = l$*. Let us introduce the interval* $[0, T_1)$*, where* $T_1 > 0$ *in uniquely defined from the condition* $\int_0^{T_1} m(s)\,ds = l$*. Then Cauchy problem (9.4.3) has positive solution* $\hat{x}(t) \in \mathcal{C}_{[0,T_1)}$*, sequence* $\{x_n(t)\}$ *(defined in (9.4.4)) converges to* $\hat{x}(t)$ *as* $n \to \infty$*,* $\lim\limits_{t \to T_1} \hat{x}(t) = \infty$*.*

Proof of existence of the desired function $\hat{x}(t)$ on the interval $[0, T_1)$ follows from proved existence of inverse mapping $\Phi^{-1} : [0, l) \to [0, +\infty)$. Since $l = \int_0^\infty \frac{dx}{\gamma(x)} = \int_0^{T_1} m(s)\,ds$, based on Lemma 2 we have $\lim\limits_{t \to T_1 - 0} \hat{x}(t) = +\infty$.

Theorem 9.3. *Let conditions* **(A)**, **(B)** *and* $\lim\limits_{x \to +\infty} \int_0^x \frac{dx}{\gamma(x)} = +\infty$ *are satisfied. Then integral equation (9.4.1) has main solution* $x(t)$*, defined for*

$t \in [0, \infty)$ *and the following a priory estimate* $|x(t)| \leq \hat{x}(t)$ *is valid, where* $\hat{x}(t)$ *is solution of the Cauchy problem (9.4.3).*

Proof. Let us introduce the sequences

$$x_n(t) = \int_0^t K(t, s, x_{n-1}(s)) \, ds, \ x_0(t) \equiv 0,$$

$$\hat{x}_n(t) = \int_0^t m(s) \gamma(\hat{x}_{n-1}(s)) \, ds, \ \hat{x}_0(t) \equiv 0.$$

Then because of the Theorem conditions we have the following inequality $|x_{n+p}(t) - x_n(t)| \leq \hat{x}_{n+p}(t) - \hat{x}_n(t)$, $|x_n(t)| \leq \hat{x}_n(t)$ for $t \in [0, T_1], T_1 < \infty$. The positive monotonic increasing sequence $\hat{x}_n(t)$ is the Cauchy sequence (it follows from the Lemma) in the norm of the space $\mathcal{C}_{[0,T_1]}$, i.e., $\max\limits_{0 \leq t \leq T_1} (\hat{x}_{n+p}(t) - \hat{x}_n(t)) \leq \varepsilon$ for $n \geq N(\varepsilon)$ and for arbitrary p. Hence, $\max\limits_{0 \leq t \leq T_1} |x_{n+p}(t) - x_n(t)| \leq \varepsilon$ for $n \geq N(\varepsilon)$, $\forall p$. Hence the sequence $\{x_n(t)\}$ is in the sphere $S(0, \hat{x}(t))$ and it is the Cauchy sequence. Because of completeness of the space $\mathcal{C}_{[0,T_1]} \ \exists \ \lim\limits_{n \to \infty} x_n(t) = x(t)$. And $|x(t)| \leq \hat{x}(t)$. Since $K(t, s, x)$ is continuous wrt x then function $x(t)$ satisfies the condition (9.4.1). The theorem is proved.

As footnote let us outline that the classic Hartman-Wintner theorem on the Cauchy problem solution continuation on semi-axis follows from this theorem. $\qquad \square$

Corollary 9.1. *Let in condition (B)* $\gamma(x) = a + bx$, $a \geq 0, b > 0$. *Then for main solution of equation (9.4.1) the following a priory estimate*

$$|x(t)| \leq a \int_0^t m(z) \exp \left(b \int_z^t m(s) \, ds \right) \, dz$$

is satisfied for $0 \leq t < \infty$.

In order to prove this corollary it is enough to verify that the Cauchy problem $\frac{dx}{dt} = m(t)(a + bx(t))$, $x|_{t=0} = 0$ has the solution

$$\hat{x}(t) = a \int_0^t m(z) \exp \left(b \int_z^t m(s) \, ds \right) \, dz$$

for $0 \leq t < \infty$.

Theorem 9.4. *Let conditions (A) and (B) be satisfied. Let* $\lim\limits_{x\to\infty} \int_0^x \frac{dx}{\gamma(x)} = l$. *Introduce the interval* $[0, T_1]$, *where* $T_1 > 0$ *is uniquely defined from the equality* $\int_0^{T_1} m(s)\,ds = l$. *Then the integral equation (9.4.1) in* $C_{[0,T_1)}$ *has the main solution* $x(t)$. *For* $t \in [0, T_1)$ *a priory estimate* $|x(t)| \leq \hat{x}(t)$, *is satisfied, where* $\hat{x}(t)$ *is the solution of the Cauchy problem (9.4.3),* $\lim\limits_{t\to T_1} \hat{x}(t) = +\infty.$

Proof follows from the Theorem 9.1 taking into account the Lemma 9.2 results.

Corollary 9.2. *(Alternative global solvability of equation (9.4.1)) Let conditions of Theorem 9.4 be fulfilled. Then either solution of the equation (9.4.1) can be continued on the whole semi-axis or on* $[T_1, +\infty)$ *or there the blow-up point exists.*

Corollary 9.3. *Let conditions of Theorem 9.2 be satisfied, and in addition let* $\gamma(x) = ax^2 + bx + c$, *where* $4ac - b^2 > 0$, $a > 0, b > 0, c \geq 0$. *Then main solution* $x(t)$ *of equation (9.4.1) exists on* $[0, T_1]$, *where positive* T_1 *is defined from condition*

$$\int_0^{T_1} m(s)\,ds = \frac{2}{\sqrt{4ac-b^2}}\left(\frac{\pi}{2} - \text{arctg}\frac{b}{\sqrt{4ac-b^2}}\right).$$

For $t \in [0, T_1)$ *we have the estimate* $|x(t)| \leq \hat{x}(t)$, *where function* $\hat{x}(t)$ *is defined by formula*

$$\hat{x}(t) = \frac{\sqrt{4ac-b^2}}{2a}\,\text{tg}\left[\text{arctg}\frac{b}{\sqrt{4ac-b^2}}\right.$$

$$\left. + \frac{\sqrt{4ac-b^2}}{2}\int_0^t m(s)\,ds\right] - \frac{b}{2a}. \tag{9.4.5}$$

Proof. Solution to the corresponding Cauchy problem (9.4.3) can be constructed easily when Corollary 9.3 conditions are satisfied. Indeed, $\Phi(x) = \int_0^x \frac{dx}{ax^2+bx+c} = \frac{2}{\sqrt{4ac-b^2}}\left(\text{arctg}\frac{2ax+b}{\sqrt{4ac-b^2}} - \text{arctg}\frac{b}{\sqrt{4ac-b^2}}\right)$, $M(t) = \int_0^t m(s)\,ds$. As results the equality $\int_0^\infty \frac{dx}{\gamma(x)} = \int_0^{T_1} m(s)\,ds$, which serves to find T_1, means that

$$\frac{2}{\sqrt{4ac-b^2}}\left(\frac{\pi}{2}-\mathrm{arctg}\frac{b}{\sqrt{4ac-b^2}}\right)=\int_0^{T_1}m(s)\,ds. \qquad (9.4.6)$$

From equation $\Phi(x)=M(t)$ it follows that majorizing function $\hat{x}(t)$, on $[0,T_1)$, must be constructed with formula (9.4.5). ∎

This Corollary can be used in the problem of extending the solution of the equation (9.4.1) with parameters. Indeed, let conditions of Theorem 9.2 be satisfied and function K depends on parameter $\lambda \in \mathbb{R}^n$, $||\lambda|| < \delta$ (i.e., $K = K(t,s,x,\lambda)$). Then in condition **B** function $\gamma(|x|,||\lambda||)$ will depend on this parameter's norm. Let $\gamma(|x|,0)=0$. Then equation (9.4.1) has trivial solution for $\lambda=0$. If $\lambda\neq 0$ equation (9.4.1) can has nontrivial main solution. Next result allows us to estimate the interval where exists main solution to the equation (9.4.1) for $0<||\lambda||<\delta$.

Let conditions of Theorem 9.2 are satisfied and let $m(s)=1$, $\gamma(x,\lambda)=a(||\lambda||)x^2+b(||\lambda||)x+c(||\lambda||)$. Let $a(||\lambda||)$, $b(||\lambda||)$, and $c(||\lambda||)$ are positive infinitesimal functions for $||\lambda||\to 0$. Suppose that in a punctured neighborhood $0<||\lambda||<\delta$ the following inequalities are satisfied $4a(||\lambda||)c(||\lambda||)-b^2(||\lambda||)=\Delta(||\lambda||)>0$,

$$\sup_{||\lambda||<\delta}\arctan\frac{b(||\lambda||)}{\sqrt{\Delta(||\lambda||)}}=\sigma,\ \sigma<\frac{\pi}{2}.$$

Then for $0<||\lambda||<\delta$ equation (9.4.1) has the main solution $x(t,\lambda)$, defined on the intervals $[0,T(||\lambda||)$, $T(||\lambda||)=\frac{2}{\sqrt{\Delta(||\lambda||)}}\left(\frac{\pi}{2}-\sigma\right).$ Therefore $\lim_{||\lambda||\to 0}T(||\lambda||)=+\infty$ and the following a proper estimate for the main solution is satisfied $|x(t,\lambda)|\leq\hat{x}(t,\lambda)$. Here the positive majorizing function $\hat{x}(t,\lambda)$ is defined on the interval $[0,T(||\lambda||)$ using formula (9.4.5) where coefficients are defined as follows: $a=a(||\lambda||),b=b(||\lambda||)$, $c=c(||\lambda||)$, $\lim_{t\to T(||\lambda||)}\hat{x}(t,\lambda)=+\infty.$

Algebraic majorants

In order to estimate the guaranteed closed interval $[0,T]$ for existence of main solution of the equation (9.4.1) and its norm estimation $\mathcal{C}_{[0,T]}$ in the algebraic majorants are useful.

Indeed, let condition (A) be satisfied and let in addition the following condition be satisfied:

(A) Let there exists continuous, differentiable and convex w.r.t. r, positive and monotonic increasing function $M(\rho, s, r)$, defined for $\rho \geq 0, s \geq 0, r > 0$ such as in the area D for $t \in [0, \rho], |x| \leq r$ the following inequalities are satisfied

$$\left| \int_0^t K(t, s, x(s)) \, ds \right| \leq \int_0^\rho M(\rho, s, r) \, ds,$$

$$\left| \int_0^t K'(t, s, x(s)) \, ds \right| \leq \int_0^\rho M'_r(\rho, s, r) \, ds.$$

Let us introduce function

$$M(\rho, r) = \int_0^\rho M(\rho, s, r) \, ds$$

and it's positive derivative

$$M'_r(\rho, r) = \int_0^\rho M'_r(\rho, s, r) \, ds.$$

Lemma 9.6. *System*

$$\begin{cases} r = M(r, \rho), \\ 1 = M'_r(r, \rho) \end{cases} \tag{9.4.7}$$

has unique positive solution r^*, ρ^*. Moreover, for any $\rho \in [0, \rho^*]$ the equation $r = M(r, \rho)$ has main solution $r(\rho)$, i.e., monotonic increasing sequence $r_n = M(r_{n-1}, \rho), r_0 = 0$, converges to the solution of equation $r = M(r, \rho)$ for any $\rho \in [0, \rho^*]$.

Proof is geometrically obvious, if one consider the graphs of the curves $y = M(r, \rho)$ for various ρ and bisection $y = r$ on the plane (y, r). Line $y = r$ tangents the curve $y = M(r, \rho^*)$ in the point (r^*, ρ^*),

Theorem 9.5. *Let conditions (A) and (C) be satisfied, (r^*, ρ^*) is positive solution of system (9.4.7). Then main solution $x(t)$ of equation (9.4.1) exists in $\mathcal{C}_{[0, \rho^*]}$ and the following estimate is satisfied*

$$\max_{0 \leq t \leq \rho^*} |x(t)| \leq r^*.$$

Proof. Let us introduce two sequences

$$x_n(t) = \int_0^t K(t, s, x_{n-1}(s))\, ds,$$

$$r_n = M(r_{n-1}, \rho^*),$$

where $x_0(t) = 0$, $r_0 = 0$. Then the following inequality $||x_{n+p} - x_n||_{C_{[0,\rho^*]}} \le r_{n+p} - r_n$ is valid for $n \ge N(\varepsilon)$ and arbitrary p. Therefore since due to Lemma 9.6 we have $\lim_{n \to 0} r_n = r^*$, then $r_{n+p} - r_n \le \varepsilon$ for $n \ge N(\varepsilon)$ and for arbitrary p. Therefore the sequence $\{x_n(t)\}$, for $x_0 = 0$ remains fundamental, $||x_n|| \le r^*$ and the theorem is proved.   □

Let us consider the following example:

$$x(t) = \int_0^t \left(K_2(t, s)x^2(s) + K_1(t, s)x(s) + K_0(t, s) \right) s^2\, ds.$$

Let

$$\sup_{0 \le s \le t < \infty,\ i=1,2,3} |K_i(t, s)| \le 1.$$

The corresponding majorant algebraic system

$$\begin{cases} r = \dfrac{\rho^3}{3}(1 + r + r^2), \\[2mm] 1 = \dfrac{\rho^3}{3}(1 + 2r) \end{cases}$$

has the following solution: $r^* = 1$, $\rho^* = 1$. Therefore, based on Theorem 9.3 the integral equation has the main solution $x(t) \in C_{[0,1]}$, $||x|| \le 1$. The integral majorant $x(t) = \int_0^t (x(s)^2 + x(s))\, ds + \frac{t^3}{3}$ and Corollary 9.3 gives us more complete information regarding the solution. Indeed, the integral equation has continuous solution $x(t)$ on the interval $[0, 1.5365)$ and the following estimate is satisfied $|x(t)| \le 0.8660 \tan \left(0.5236 + 0.2886t^3 \right) - 0.5$ for $0 \le t < 1.5365$.

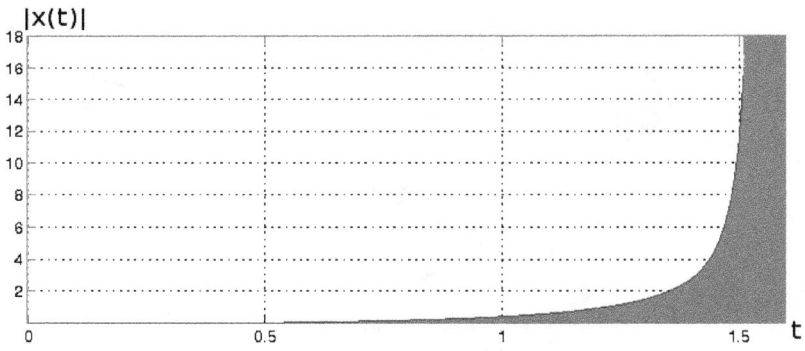

Fig. 9.1.   Majorant of the solution. Solution $|x(t)|$ belongs to the gray zone.

Chapter 10

Generalized Solutions to Nonlinear Volterra Equations of the First Kind

10.1 Problem Statement

In this chapter we describe the algorithm for construction of the generalized solutions to the following nonlinear integral equation

$$\int_0^t K(t,s)(x(s) + g(s^l x(s), s))ds = f(t), \qquad (10.1.1)$$

where K, f and g are analytic functions in a neighborhood of zero, moreover the kernel K enjoying the property

$$K(t,s) = \sum_{i=0}^n K_{n-i}^n t^{n-i} s^i + \mathcal{O}((|t| + |s|)^{n+1}), \quad K_{n-i}^n \neq 0,$$

$$i = 0, 1, \ldots, n, n < l.$$

The objective of the present chapter is to construct generalized solutions

$$x(t) = c_0\delta(t) + c_1\delta^{(1)}(t) + \cdots + c_0\delta^{(n)}(t) + u(t), \qquad (10.1.2)$$

where $\delta(t)$ is the Dirac delta function (or, more correctly, functional), and $u(t)$ is a regular function.

Outline

This chapter is organized as follows. First, in Section 10.2 we reduce the problem of solution's parameters $c_0, \ldots c_1$ estimation to the special system of linear algebraic equations with lower triangular matrix. By this means we construct the singular part of the solution. In Section 10.3 we employ combination of the method of undetermined coefficients with the successive approximation method to find the regular part of the desired solution to the equation (10.1.1). Finally in Section 10.4 we provide an example to illustrate the developed method.

10.2   Singular Component Determination

The set of all infinitely differentiable compactly supported functions with supports in the neighborhood $(-\rho, \rho)$ is denoted by $D_{(-\rho,\rho)}$. The set of linear continuous functionals defined on $D_{(-\rho,\rho)}$ is denoted by $D'_{(-\rho,\rho)}$ and the subset of elements of the form (10.1.2) with an nth order singularity supported at zero is denoted by $D'_{n(-\rho,\rho)}$. Thus the solution (10.1.2) of (10.1.1) is sought in the class $D'_{n(-\rho,\rho)}$ and should satisfy the equation (10.1.1) in the sense of the Sobolev-Schwartz distributions theory, readers may refer to classic textbooks of [Vladimirov (1979)] and [Kanwal (2013)]. Note that the product $t^l x = t^l u(t)$ is a regular function for $n < l$ and for all $x \in D'_{n(-\rho,\rho)}$, which provides a solution of the problem of nonlinear operations with such distributions for equation (10.1.1) with $l > n$. Since the identities

$$t^{k-i}\Theta * s^i \delta^{(j)}(s) = (-1)^J j! t^{k-j} \delta$$

are valid in the space D' for $i, j = 0, 1, ..., n, k \geq n$, where Θ is the Heaviside function and δ_{ij} is the Kronecker delta, we have the relation

$$\int_0^t \sum_{k=n}^{\infty} \sum_{i=0}^{k} K_{k-i}^k t^{k-i} s^i (c_0 \delta(s)$$

$$+ \ldots + c_n \delta^{(n)}(s)) \, ds = \sum_{j=0}^{n} (-1)^j j! \sum_{k=n}^{\infty} K_{k-j}^k t^{k-j} c_j.$$

Let us note the following equality

$$\sum_{j=0}^{n} (-1)^j \left. \frac{\partial^j K(t, s)}{\partial s^j} \right|_{s=0} = \sum_{j=0}^{n} (-1)^j j! \sum_{k=n}^{\infty} K_{k-j}^k t^{k-j}.$$

Therefore, an element $x \in D'_{n(-\rho,\rho)}$ can be a solution of equation (10.1.1) only if the regular component in the representation (10.1.2) satisfies the following equation

$$\int_0^t K(t, s)(u(s) + g(s^l u(s), s)) ds = r(t, c_0, \ldots, c_n), \qquad (10.2.1)$$

where

$$r(t, c_0, \ldots, c_0) = f(t) - \sum_{j=0}^{n} (-1)^j \frac{\partial^j K(t, 0)}{\partial s^j} c_j.$$

We find the parameters c_j from the following SLAE

$$r_t^{(i)}(0, c_0, \ldots, c_n) = 0, i = \overline{0, n}, \qquad (10.2.2)$$

with lower triangular matrix with diagonal entries $K_0^n, K_1^n, \ldots, K_n^n$.

If these numbers are nonzero, then the constants c_j can be found successively and uniquely. If some of the diagonal entries are zero and the vector $\{f(0), f'(0), \ldots, f^{(n)}(0)\}'$ satisfies the resolvability condition, then part of of the constants on the right-hand side in (10.2.1) can remain arbitrary.

If $f^{(i)}(0) = 0$, $K_i^n = 0, i = \overline{0, k-1}$ and $K_i^n \neq 0$ for $i = \overline{k, n}$, then, by setting $c_k = \cdots = c_n = 0$, one can uniquely find the constants $c_0, \ldots, c_{k-1}$ from the system (10.2.2).

10.3   Regular Component Determination

To construct the regular function $u(t)$, we solve equation (10.2.1) for already known values $c_0, \ldots, c_n$ by combining the method of undetermined coefficients with the successive approximation method. For sake of clarity let us introduce the following notation

$$\Phi(u, t) := \int_0^t \sum_{i=0}^{n} K(t, s)(u(s) + g(s^l u(s), s)) \, ds - r(t, c) = 0. \qquad (10.3.1)$$

Suppose that the homogeneous equation

$$\sum_{i=0}^{t} \sum_{i=0}^{n} K_{n-i}^n t^{n-i} s^i x(s) ds = 0 \qquad (10.3.2)$$

possesses only the trivial solution. This is the case if $\sum_{i=0}^{n} K_{n-i}^n (i+j)^{-1} \neq 0$ for $j = 1, 2, \ldots$ then for each positive integer N, there exist constants u_i such that

$$\left| \Phi(u_0 + u_1 t + \cdots + u) N t^N, t) \right| = \mathcal{O}(|t|^{n+N+1}). \qquad (10.3.3)$$

Since the homogeneous equation (10.3.2) possesses only the trivial solution, we have

$$\int_0^t \sum_{i=0}^{n} K_{n-i}^n t^{n-i} s^{i+j} ds \neq 0$$

for $j = 0, 1, 2, \ldots$, and the coefficients u_i are uniquely determined by means of the method of undetermined coefficients after the substitution of the polynomial

$$u^0(t) = u_0 + u_1 t + \cdots + u_N t^N$$

into the equation (10.3.1).

Next, we substitute the function

$$u(y) = u^0(t) + t^N v(t) \tag{10.3.4}$$

into equation (10.3.1) and collect terms containing the powers t^i, $i = n, n + 1, \ldots, n + N$, taking into account the relation (10.2.2) and the definition of the polynomial $u^0(t)$. Then we differentiate the resulting relation w.r.t. t and find the function v by the successive approximation method from the following integral equation

$$v = F(v, t). \tag{10.3.5}$$

Here

$$F(v, t) = \frac{1}{K(t, t) t^N} \left\{ -K(t, t)(u^0(t) + g(t^l u^0(t) + t^{l+N} v(t), t) \right.$$

$$- \int_0^t K_t'(t, s)(u^0(s) + s^N v(s) + g(s^l u^0(s)$$

$$\left. + s^{l+N} v(s), s)) \, ds + r_t'(t, c) \right\}.$$

Suppose that

$$\sum_{i=0}^n K_{n-i}^n = a \neq 0.$$

Let us show that the operator F satisfies the assumptions of the contraction mapping principle in the ball $||x|| \leq r$ of the space $\mathcal{C}_{[-\rho, \rho]}$ for sufficiently large N. Indeed,

$$|g(s^l(u^0(s) + s^N v_1(s)), s) - g(s^l(u^0(s) + s^N v_2(s)), s)|$$

$$\leq |s|^{l+N} C_1 |v_1 - v_2|$$

for all v_1 and v_2 in the ball $S(0, r) \in \mathcal{C}_{[-\rho, \rho]}$. Since

$$|K_t'(t, s)| \leq C_2(|t| + |s|)^{n-1},$$

we have

$$\left| \frac{1}{t^{n+N}} \int_0^t K_t'(t, s) s^N \, ds \right| \leq \frac{2^{n-1} C_2}{N + 1}.$$

By virtue of these bounds, there exists a constant c such that
$$|F(v_1,t) - F(v_2,t)| \le \frac{c}{N+1}||v_1 - v_2||.$$
We next fix a $q < 1$ and take $N > c/q - 1$. Then F is a contractive operator with exponent q in the ball $||v|| \le r$ of the space $C_{[-\rho,\rho]}$. Since by (10.3.3), $|F(0,t)| = \mathcal{O}(|t|)$, it follows that there exists a $\hat\rho \in (0, \rho]$ such that
$$\max_{|t| \le \hat\rho} |F(0,t)| \le (1-q)r.$$
Consequently, the contraction operator F maps the ball $|v| \le r$ of the space $C_{[-\hat\rho,\hat\rho]}$ into itself. This implies the following assertion.

Theorem 10.1. *Let $k > n$, and let*
$$\sum_{i=0}^{n} K_{n-1}^{n} \frac{1}{i+j} \ne 0, j = 1,2,\ldots,$$
$$K_{n-i}^{n} \ne 0, i = 0,1,\ldots,n, \sum_{i=0}^{n} K_{n-i}^{n} K_{n-i}^{n} \ne 0.$$
Then equation (10.1.1) *possesses a unique solution* (10.1.2), (10.3.4) *in the class $D_n'(-\hat\rho,\hat\rho)$, where the constants c_j are computed by the method of undeterminated coefficients from equation* (10.3.1), *and the continuous function $v(t)$ is constructed by the successive approximations method from equation* (10.3.5).

It is to be noted that instead of analyticity of K, g and f in the Theorem 10.1, one could only require that these functions are sufficiently smooth. In the analytical case, the entire regular part $u(t)$ of the solution (10.1.2) is an analytical function in a neighborhood of zero, and its Taylor coefficients can be identified by means of the method of undetermined coefficients from equation (10.3.1). If $f^{(i)}(0) = 0, i = \overline{0,n}$, then all $c_i = 0$ in the solution (10.1.2), and this solution is classical. If, under assumption of Theorem 10.1, some of the elements $K_i^n, i = \overline{0,n}$ are zeros and, in addition, system (10.2.2) is solvable, then the solution (10.1.2), (10.3.4) depends on k arbitrary constants, where $k = n + 1 - r$ and r is the rank of the matrix of the linear system (10.2.2).

Theorem 10.1 can be strengthened as follows

Theorem 10.2 ([Sidorov and Sidorov (2006)]). *Under the assumptions of Theorem 10.1, let $\sum_{i=0}^{n} K_{n-i}^{n} = 0$ and moreover, let*
$$\left.\frac{\partial^i K(t,s)}{\partial t^i}\right|_{s=t} = 0, i = 0,1,\ldots,p-1,$$

$$\left.\frac{\partial^p K(t,s)}{\partial t^p}\right|_{s=t} = \mathcal{O}(t^{n-p}), \; p \le n.$$

Then the assertion of Theorem 10.1 remains valid.

10.4   Example

Let us consider the following equation

$$\int_0^t (t^2 + ts - 2s^2)(x(s) + s^5 x^2(s)) \, ds = 1 + t + t^2 + t^3.$$

Here the assumptions of Theorem 10.2 are valid for $n = 2$, $l = 5/2$, and $p = 1$. In the class D' our equation has the following solution

$$x(t) = \delta(t) - \delta^{(1)}(t) - \frac{1}{4}\delta^{(2)}(t) + \frac{-1 + \sqrt{1 + t^5 \times 25/5}}{2t^5},$$

those singular part is determined by (10.2.2) and whose regular part has been found from the equation

$$\int_0^t (t^2 + ts - 2s^2)(u(s) + s^5 u^2(s)) \, ds = t^3. \qquad (10.4.1)$$

This equation has the following analytic solution

$$u(t) = \frac{-1 + \sqrt{1 + t^5 \times 24/5}}{2t^5}.$$

The Taylor coefficients pf the solution at the point $t = 0$ can be computed by means of the method of undetermined coefficients. Note that, along with this solution, equation (10.4.1) has the solution

$$u_2(t) = \frac{-1 - \sqrt{1 + t^5 \times 24/5}}{2t^5}$$

for which the point $t = 0$ is a fifth-order pole. In general case equation (10.1.1) may possesses a several branching solutions. Such solutions can be constructed with the use of the results of this chapter in combination with results presented in the book of [Vainberg and Trenogin (1974)].

Chapter 11

On Impulse Control of Nonlinear Dynamical Systems Based on the Volterra Series

11.1 Introduction. The State of the Art

The Volterra series characterizes a system as a mapping between two function spaces, the input and output spaces of that system. The Volterra
model is an extension of the Taylor series representation to cover the nonlinear dynamical systems and known as one of the major nonlinear system
modelling tool used for process control. Here readers may refer to books
of [Schetzen (1980); Rugh (1981); Corduneanu (2002)]. The series can be
described in the time-domain or in the frequency-domain. The Volterra
functional series have been employed in many fields in electrical and power
engineering during the last decades (here readers may refer to the results of
[Chua and Tang (1982); Wiener and Spina (1980); Larsen (1993); Chua and
Liao (1991); Shcherbakov (1996); Sidorov (1999); Scherbinin (2010)]). It is
easy to see that the Volterra series appears to be a strong tool in processing
the nonlinear problems. In behavioral (or black box) modelling of input–
output nonlinear systems with memory the following integral-functional
Volterra series

$$\sum_{m=1}^{\infty} F_m = f(t), \ 0 \le t \le T \tag{11.1.1}$$

where

$$F_m := \int_0^t \ldots \int_0^t K_m(t, s_1, \ldots, s_m) \prod_{i=1}^m x(s_i) ds_i$$

is one of the most common technique (see the bibliography in the publications of [Rugh (1981); Chua and Tang (1982); Chua and Liao (1991);
Apartsyn (2003)]). In the Volterra model (11.1.1) the signal $f(t)$ is the

response of the system, $x(t)$ is the input signal, functions $K_m(t, s_1, \ldots, s_m)$ are time-varying transfer functions or Volterra kernels as shown in Fig. 11.1.

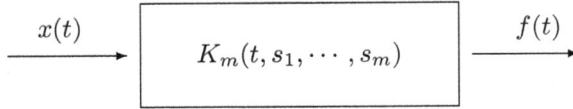

$$x(t) \longrightarrow \boxed{K_m(t, s_1, \cdots, s_m)} \longrightarrow f(t)$$

Fig. 11.1. Black box system.

Convergence of the Volterra series follows from the following classical Fréchet Theorem 11.1. Here readers may also refer to the classical monograph of [Volterra (2005)].

Theorem 11.1 ([Fréchet (1910)]). *Any continuous functional $G[y(t)]$ can be represented as*

$$G[y(t)]_{t \in [a,b]} = \lim_{n \to \infty} G_{r_n}[y(t)],$$

where

$$
\begin{aligned}
G[y(t)]_{t \in [a,b]} = \lim_{n \to \infty} \Bigg[& k_{n,o} + \int_a^b k_{n,1}(\xi) y(\xi) d\xi \\
& + \int_a^b \int_a^b k_{n,2}(\xi_1, \xi_2) y(\xi_1) y(\xi_2) d\xi_1 d\xi_2 + \cdots \\
& + \int_a^b \int_a^b \cdots \int_a^b k_{n,r_n}(\xi_1, \xi_2, \cdots, \xi_{r_n}) y(\xi_1), \cdots, \\
& \times y(\xi_{r_n}) d\xi_1 d\xi_2 \cdots d\xi_{r_n} \Bigg],
\end{aligned}
$$

where $k_{n,r_n}(\xi_1, \cdots, \xi_{r_n})$ are continuous functions defined for the functional G regardless of the function $y(t)$.

Theorem 11.1 generalizes the famous Weierstrass approximation theorem which states that every continuous function defined on a closed interval can be uniformly approximated as closely as desired by a polynomial function. The book of [Volterra (2005)] demonstrates that such series converges uniformly in any compact set of continuous functions, for any $y \in K$, where K is compact set of continuous functions, and $\left| G[y(t)] - G_{r_n}[y(t)] \right| < \varepsilon$ where $r_n \geq N(\varepsilon)$. Here readers may also refer to article of [Baesler and

Daugavet (1990)]. The overview of nonlinear operator approximation theory and computational methods is presented in the monograph of [Torokhti and Howlett (2007)].

It is to be noted that finite models obtained from (11.1.1) are applicable for single input single output (SISO) dynamical systems only. In case of multiple input single output models (MISO) dynamical systems one may employ the following generalized Volterra model

$$\sum_{p=1}^{q} \sum_{1 \leq i_1 \leq \ldots \leq i_p \leq m} \int_0^t \cdots \int_0^t K_{i_1 \ldots i_p}(t, s_1, \ldots, s_p) x_{i_1}(s_1) \ldots x_{i_p}(s_p) \, ds_1 \ldots ds_p$$

$$= y(t), \tag{11.1.2}$$

where $t \in [0, T]$. Here transfer functions $K_{i_1 \ldots i_p}(t, s_1, \ldots, s_p)$ are already not assumed to be symmetric w.r.t the variables $s_1, \ldots, s_p$. For more theory and numerical methods related to MISO non-stationary Volterra models identification readers may refer to results of [Sidorov (1999, 2002)].

It is to be outlined that such *behavioral* or *black box models* of nonlinear dynamical systems (i.e., in the absence of a detailed description of the system structure) is important for the experiment-based characterization and performance evaluation of complex nonlinear systems. Volterra models are also directly related to *polynomial kernel regression* and support vector regression and can be considered in the framework of *machine learning* theory. Here readers may refer to papers of [Franz and Schölkopf (2006)] and [Smola and Schölkopf (2004)]. Support vector regression and its application to nonlinear time series analysis is addressed in chapter 13.

The operator located on the left hand side of the equality (11.1.2) can be considered as the special case of Lyapunov-Schmidt integral operator, employed by [Lyapunov (1906)] and later by [Lichtenstein (1931)] in classic studies of mechanics. The predictive control algorithms based on a Volterra model are considered by [Gruber *et al.* (2011)].

If a system to be modeled is stationary in the sense that its dynamical characteristics during the transient process T can be considered as constant, the Volterra kernels K_m depend only on the differences $t - s_i$, $i = \overline{1, m}$ and then instead of (11.1.1) the following model can be employed

$$\sum_{m=1}^{N} \int_0^t \cdots \int_0^t K_m(s_1, \ldots, s_m) \prod_{i=1}^{m} x(t - s_i) \, ds_i$$

$$= f(t), \quad t \in [0, T]. \tag{11.1.3}$$

Here the kernels K_m are continuous and symmetric with respect to a set of variables.

Outline

The chapter is organized as follows. First we provide the review of the Volterra models identification methods in Section 11.2. We outline the key methods for the Volterra models identification from process input-output data since the Norbert Wiener's works on the Volterra series based models in 1950s. In Section 11.3 we concentrate on impulse control construction as solution of polynomial (or multiple) Volterra integral equations of the first kind. Section 11.4 outlines two examples. Finally some summarizing remarks are presented in the conclusion.

11.2 Overview of the Volterra Models Identification Methods

For $N = 1$ in (11.1.3) we have the output signal representation with a convolution Duhamel integral

$$\int_0^t K_1(s) \, x(t - s) \, ds = y(t), \, t \in [0, T],$$

which is a conventional model in the linear theory of automatic control. In this case for identification of $K_1(t)$ it is sufficient to know a response of the system to certain signal. If, in particular, one assume $x(t) = \alpha \theta(t)$, where $\theta(t)$ is the Heaviside function, and $\alpha \neq 0$ a scalar, then at $y(t) \in C_{[0,T]}^{(1)}$, $y(0) = 0$, we have the following inversion formula $K_1(t) = \frac{y'(t)}{\alpha}$. The same method can be generalized and applied for identification of the transfer functions in the Volterra model (11.1.3) of dynamical systems with stationary transfer functions. Indeed, for the identification of transfer functions K_m in (11.1.3) the following $(m-1)$ – parametric family of piecewise constant training signals

$$x_{\omega_1,\dots,\omega_{m-1}}^{\alpha_k}(t) = \alpha_k \sum_{i=0}^{m-1} \gamma_i \theta(t - \beta_i) \tag{11.2.1}$$

can be employed. Here readers may refer to the papers of [Apartsyn (2003); Apartsyn and Solodusha (2004)]. Here $\gamma_i = (-1)^i \cdot 2$, $i = \overline{1, m-2}$, $\gamma_0 = 1$, $\gamma_{m-1} = (-1)^{m-1}$, $\beta_i = \sum_{j=1}^i \omega_j$, $\beta_0 = 0$, $\omega_1, \dots,$ ω_{m-1}, $t \in [0, T]$, $k = \overline{1, m}$, $\alpha_1 \neq \alpha_2 \neq \cdots \neq \alpha_m \neq 0$. In particular for the practical cases $m = 2, 3$ we have

$$x_{\omega_1}^{\alpha_k}(t) = \alpha_k \left[\theta(t) - \theta(t - \omega_1) \right], \, k = 1, 2;$$

$$x^{\alpha_k}_{\omega_1,\omega_2}(t) = \alpha_k \left[\theta(t) - 2\theta(t - \omega_1) + \theta(t - \omega_1 - \omega_2) \right], \quad k = 1, 2, 3.$$

Let us denote by $f_m(t, \omega_1, \dots, \omega_{m-1})$ (see also below) a component of the system responses to the parametric family of piecewise constant signals (11.2.1) that reflects contribution of kernel K_m in the model (11.1.3), then taking into account the symmetry of K_m with respect to all the variables, its identification is reduced to solving the following linear N-dimensional Volterra equation of the first kind

$$V_m K_m \equiv \sum_{i_1 + \cdots + i_{m-1} = m} (-1)^{\sum\limits_{k=1}^{\left[\frac{m-1}{2}\right]} i_{2k}} \frac{m!}{i_1! \dots i_{m-1}!} V_{i_1, \dots, i_{m-1}} K_m = f_m,$$

where

$$V_{i_1, \dots, i_{m-1}} K_m$$

$$= \overbrace{\underbrace{\int \cdots \int}_{\substack{t - \omega_1 \\ i_1 \text{ times}}}}^{t} \cdots \overbrace{\underbrace{\int \cdots \int}_{\substack{t - \omega_1 - \cdots - \omega_{m-1} \\ i_{m-1} \text{ times}}}}^{t - \omega_1 - \cdots - \omega_{m-2}} K_m(s_1, \dots, s_m) \, ds_1 \dots ds_m,$$

$$f_m \equiv f_m(t, \omega_1, \dots, \omega_{m-1}),$$

$$t, \omega_1, \dots, \omega_{m-1} \in \triangle_m = \left\{ t, \omega_1, \dots, \omega_{m-1} / 0 \le \sum_{k=1}^{m-1} \omega_k \le t \le T \right\},$$

and the symbol $[\cdot]$ in the limit superior of summation means an integer part of the value. For $m = 2$ we have

$$V_2 K_2 \equiv \int_{t - \omega_1}^{t} \int_{t - \omega_1}^{t} K_2(s_1, s_2) \, ds_1 ds_2 = f_2(t, \omega_1).$$

It's easy to see that the inversion formula is as follows:

$$K_2(t, t - \omega_1) = \frac{1}{2} \left(\frac{\partial^2 f_2(t, \omega_1, \omega_2)}{\partial t \partial \omega_1} + \frac{\partial^2 f_2(t, \omega_1, \omega_2)}{\partial \omega_1^2} \right).$$

It is to be noted that the practical significance of these formulas is not high due to the instability of the operation of numerical differentiation of empiric functions. If we replace the variables $\omega_i = s_i - s_{i+1} \ge 0$, $i = \overline{1, m-1}$, $s_1 = t$, then, (here readers may also refer to the monograph of [Apartsyn (2003)]) the inversion formula can be represented as follows

$$K_m(s_1, \dots, s_m) = \frac{(-1)^{\left[\frac{m}{2}\right]}}{m! 2^{m-2}} f^{(m)}_{m_{s_1, \dots, s_m}}(s_1, s_1 - s_2, \dots, s_{m-1} - s_m),$$

where $f^{(m)}_{m_{s_1},\ldots,s_m}$ stands for partial derivative of f_m w.r.t. $s_1,\ldots,s_m$.

Let us briefly address the non-stationary Volterra model (11.1.1) (see also thesis of [Sidorov (1999)]) and for sake of clarity let us concentrate only on single term of the series (11.1.1) corresponding to the scalar input Volterra model only:

$$\int_0^t \ldots \int_0^t K_p(t, s_1, \ldots, s_p) x(s_1) \ldots x(s_p)\, ds_1 \ldots ds_p = y(t), \qquad (11.2.2)$$

Here $t \in [0, T]$. We use the following learning signals of the family (11.2.1)

$$x_{\omega_1 \ldots \omega_p}(t) = \theta(t) + 2\sum_{k=1}^{p-1}(-1)^k \theta\left(t - \sum_{i=1}^{k}\omega_i\right) + (-1)^p\theta\left(t - \sum_{i=1}^{p}\omega_i\right). \quad (11.2.3)$$

The integration interval in (11.2.2) is splitted as follows $\Omega_{i_1,\ldots,i_p} = \{[0, \omega_1], [\omega_1, \omega_1 + \omega_2], \ldots, [\omega_1 + \cdots + \omega_{p-1}, \omega_1 + \cdots + \omega_p]\}$.

Based on the integral additivity property and based on the kernels $K_p(t, \nu_1, \cdots, \nu_p)$ symmetry w.r.t. $\nu_1, \cdots, \nu_p$, we obtain the following multidimentional Volterra equation

$$V_p K_p \equiv \sum_{i_1 + \cdots + i_p = p}(-1)^{i_2 + \cdots + i_{2[p/2]}}\frac{p!}{i_1! \cdots i_p!}V_{i_1 \ldots i_p}K_p = g_p, \quad p \geq 1,$$

$$(11.2.4)$$

where

$$g_p \equiv g_{i_1 \ldots i_p}(t, s_1, \ldots, s_p) \equiv f_{i_1 \ldots i_p}(t, \omega_1, \omega_1 + \omega_2 \ldots, \omega_1 + \cdots + \omega_p), \quad (11.2.5)$$

$$s_1 = \omega_1 + \cdots + \omega_p, \cdots, s_{p-1} = \omega_1 + \omega_2, s_p = \omega_1, \qquad (11.2.6)$$

$$V_{i_1 \ldots i_p}K_p = \underbrace{\int_0^{s_p} \cdots \int_0^{s_p}}_{i_1\,\text{times}} \cdots \underbrace{\int_{s_2}^{s_1} \cdots \int_{s_2}^{s_1}}_{i_p\,\text{times}} K_p(t, \nu_1, \ldots, \nu_p) d\nu_1 \cdots d\nu_p,$$

$$(11.2.7)$$

$$t, \omega_1, \ldots, \omega_p \in \Delta_p = \left\{t, \omega_1, \ldots, \omega_p \,/\, 0 \leq \sum_{k=1}^{p}\omega_k \leq t \leq T\,;\, \omega_k \geq 0\right\}.$$

VIE (11.2.7) has the following explicit inversion formula

$$K_p(t, s_1, \cdots, s_p) = \frac{(-1)^{[\frac{p}{2}]}}{p! 2^{p-1}}g^{(p)}_{p_{s_1, \cdots, s_p}}(t, s_1, s_2, \cdots, s_p). \qquad (11.2.8)$$

where $g^{(p)}_{p_{s_1},\ldots,s_p}$ stands for partial derivative of g_p w.r.t. $s_1, \ldots, s_p$.

In order to change the variables to $\omega_1, \cdots, \omega_p$ one can use the following transform $\bar{s} = A\bar{\omega}$, $\bar{s} = (s_1, \ldots, s_p)'$, $\bar{\omega} = (\omega_1, \ldots, \omega_p)'$ and

$\nabla_s = (A^{-1})^T \nabla_\omega$, where

$$
\nabla_s = \begin{bmatrix} \dfrac{\partial}{\partial s_1} \\ \vdots \\ \dfrac{\partial}{\partial s_p} \end{bmatrix}, \nabla_\omega = \begin{bmatrix} \dfrac{\partial}{\partial \omega_1} \\ \vdots \\ \dfrac{\partial}{\partial \omega_p} \end{bmatrix}, \quad A = \begin{bmatrix} 1 & 1 & \cdots & & 1 \\ 1 & \cdots & & 1 & 0 \\ \vdots & & \cdot & & \vdots \\ & & \cdot & & \\ 1 & 0 & \cdots & & 0 \end{bmatrix}.
$$

The problem of identification non-stationary finite Volterra models (11.1.1) (when the kernels K_m explicitly depend on time t) is briefly addressed in the next Subsection 11.2.1, also readers may refer to publications of [Sidorov (2002, 1999); Apartsyn (2003); Apartsyn and Solodusha (2004)]. The application of this approach for modelling the heat exchange processes in the high-temperature plant is presented in Part 3 of this book.

It is to be noted here despite to the advantages which follow from availability of explicit inversion formulas of the multidimensional integral Volterra equations of the first kind the developed method of identifying the Volterra kernels has a drawback. That drawback is related to quite severe conditions for the existence of solutions to these equations in the required classes of functions. Therefore, in order to model a response of system $y(t)$ to an input disturbance $x(t)$ the knowledge of the kernels K_m is in general redundant. Instead one may apply the product integration method, here readers may refer to the book of [Linz (1985)]. If we follow the product integration method and select the mesh step h then in the one-dimensional case we have

$$
\int_0^{ih} K_1(s)x(t-s)\,ds \approx \sum_{j=1}^{i} x(t_{i-j+\frac{1}{2}}) \int_{(j-1)h}^{jh} K_1(s)\,ds,
$$

where $t_{i-\frac{1}{2}} = (i - \frac{1}{2})h$, $i = \overline{1,n}$, $nh = T$. Therefore instead of identification of $\tilde{K}_m$ themselves, it is sufficient of identify elementary integrals of K_m. This approach is implemented for the case $K_2(s_1, s_2) = \psi(s_1)\psi(s_2)$, $K_3(s_1, s_2, s_3) = \varphi(s_1)\varphi(s_2)\varphi(s_3)$, $\psi(t)$, $\varphi(t) \in C_{[0,T]}$, and the explicit inversion formulas of corresponding systems of linear algebraic equations for a general case where K_2, K_3 are arbitrary continuous symmetric functions are presented in the paper of [Spiryaev (2006)].

In the papers of [Apartsyn (2013)] and [Belbas and Bulka (2010)] the finite Volterra series have been employed for control of nonlinear systems with memory (casual) and feed-back control . The control relevant problems can be reduced to the special classes of nonlinear Volterra integral equations we studied in Chapter 8. Here concentrate on impulse control of nonlinear dynamical systems.

For the construction of the Volterra models it appears the problem of the Volterra kernels identification based on the measured systems response $f(t)$ on series of special learning signals. The special numerical methods and algorithms were developed for this objective. These methods and references are available in the monograph of [Apartsyn (2003)] and have been employed in chemistry industry for power consumption optimization, results are presented in the article of [Scherbinin (2010)].

It is to be noted that the Volterra series (here readers may refer to the books of [Schetzen (1980); Bendat (1990)]), which is in principle capable of describing a very large class of nonlinear systems with memory, can be practically used only for mildly nonlinear systems (or, equivalently, for limited signal amplitude in strongly nonlinear systems), where the kernels of higher order ($m > 3$) can be neglected (see e.g., publications of [Chua and Liao (1991); Evans *et al.* (1995)]).

One of the earliest approach for the Volterra models identification was the cross-correlation method, here readers may refer to the publications of [Pearson *et al.* (1996); Wiener (1958); Lee and Schetzen (1965); Marmarelis and Marmarelis (1978); Schetzen (1980); Koh and Powers (1985); Doyle *et al.* (1995)]. A minimum $N + 1$-level input sequence is demanded to identify an N order Volterra model using the cross-correlation, as reported by [Nowak and Veen (1994)]. One of the earliest applications of the cross-correlation technique was reported by [Lee and Schetzen (1965)]. In partuclar the approach needs zero-mean and Gaussian distributed input signal. It is to be noted that such input signals has minor practical significance since it may not be practical to apply in a plant setting due to the continuum of levels employed in a Gaussian white noise (GWN) sequence. Another drawback is that the sequence was to be as long as possible to yield accurate parameter estimates. [Marmarelis and Marmarelis (1978)] showed that Constant-switching-pace Symmetric Random Sequences (CSRS) could be used for Volterra model identification instead of GWN signals. The advantage of such signals is the fact that their probability distributions can be varied to achieve the desired moment properties. It is to be noted that well known random binary sequence (RBS) is a CSRS.

[Koh and Powers (1985)] identified a second-order Volterra model based on the prediction error variance (PEV) minimization. This method has been adapted and applied in theses of [Soni and Parker (2004); Soni (2006)] for control-relevant identification of 3rd order Volterra model of nonlinear dynamics of polymerization reactor.

[Parker *et al.* (2001)] developed a plant-friendly tailored-sequence design approach for the identification of the linear and diagonal kernels of a second-order Volterra model. They proposed to employ the tailored input sequence to be symmetric, and it was significantly shorter than the sequence used in the cross-correlation technique. The identification of the kernels was also based on a PEV approach and their results have demonstrated a considerable improvement comparing with the cross-correlation technique. Taking into account the fact that the conventional Volterra models are linear in parameters, [Korenberg (1988)] identified the kernels directly as solution to the least-squares problem based on QR factorization technique. [Zhang *et al.* (1988)] employed the Korenberg's method to study the effects of model nonlinearity and model memory on the quality of the Volterra kernel estimates for a nonlinear model of the lung. They developed an data-driven approach for generic and simultaneous identification of the kernel and the memory length directly from available data. They found that an incorrect memory led to substantial distortion in the kernel estimates and neglecting higher-order kernel contributions led to a significant bias in the estimates of the lower-order kernels. [Ling and Rivera (2001)] presented a two-step algorithm in which a nonlinear auto-regressive model with exogenous inputs (ARX) was first estimated from the input-output data and then a Volterra model was generated from this ARX model. Based on this approach they managed to reduce the number of parameters required to identify the Volterra model. That is the case since the ARX model needs a lesser number of parameters to be estimated. Dodd *et al.* have employed a feature space in a reduced kernel Hilbert space (RKHS) for the Volterra model identification. The feature space consists of all possible polynomials in the input up to and including the Volterra model order. The Volterra kernels were then expressed in terms of this feature space in the RKHS. Thus a mapping was obtained from the Volterra kernels in the $\mathbb{R}^n$ space to a higher dimensional Hilbert space. The advantage of this approach is that functions in the Hilbert space can be approximated using a finite series of point observations. The Volterra kernel was then easily identified in the Hilbert space. However, this technique is applicable only for problems with relatively low model memory. For physical systems where the model memory is generally high, the dimension of the feature space can be prohibitively high. [Parker and Tummala (1992)] used the group method of data handling approach to identify a second-order Volterra model using neural networks. An artificial neural networks were employed as combination of linear and quadratic layers based on polynomial activation functions.

A set of input-output data was used to train the network by computing the necessary neural network coefficients, and finally an optimum combination of the coefficient set was used to determine the Volterra model parameters. While training the network, all possible combinations of the inputs were realized and tested for suitability. This is a problem for systems with large memory, as the polynomial activation functions may take a long time to converge. Furthermore, the authors also reported convergence problems when the inputs were not properly selected. There have been other approaches of [Liu *et al.* (1998); Aiordachioaie *et al.* (2001)] to identify Volterra models using neural networks, however they are also restricted to low memory and low order problems, and could potentially have convergence difficulties due to the polynomial activation functions. [Németh *et al.* (2002)] identified the Volterra kernel transfer function, i.e., they carried out the Fourier integral transform of the input and outputs to obtain the Volterra kernel. A multi-sine input signal was employed, which is characterized by its maximum frequency and the number of frequency components. Their analysis was restricted to second-order Volterra models and they approximated the second-order kernel space using interpolation functions. *B*-splines were used as the interpolation functions, and the authors obtained good results for the quadratic case. However, extension to higher orders seems to be difficult.

From this overview of the Volterra models identification algorithms readers may see that the main criticism in using the Volterra series as nonlinear models lies in its large number of parameters needed to be estimated. One of the ways to overcome this drawback is to approximate the Volterra kernels by certain basis functions expansions, e.g., by Laguerre expansion (see paper of [Qingsheng and Zafiriou (1995)]). They introduced the following definitions

Definition 11.1. A kernel K_m is called separable if $K_m(s_1, \ldots, s_m) = \sum_{j=1}^{q} \nu_{1j}(s_1)\nu_{2j}(s_2) \cdots \nu_{mj}(s_m)$, where q is some finite number and each $\nu_{kj}(s_k)$ is a single variable real function. If each $\nu_{kj}(s_k)$ also satisfies $\sum_{s_k=0}^{\infty} |\nu_{kj}(s_k)| < \infty$ the kernel is called stable separable.

Definition 11.2. A symmetric kernel is called strictly proper if $K_m(s_1, \ldots, s_m) = 0$ for $s_1 \cdots \cdots s_m = 0$.

For the Volterra models meeting such properties it is possible to employ the Volterra-Laguerre model. Laguerre functions (see also recent results of [Diouf *et al.* (2012)]) $\phi_{m,i}(s)$ form a complete orthonormal set in l^2 and can be conveniently defined by their z-transform as follows

$$\Phi_{m,i}(z) = \sqrt{1 - a_m^2} \frac{z}{z - a_m} \left(\frac{1 - a_m z}{z - a_m} \right)^i,$$

where $a_m \in]-1, 1]$. Therefore each Volterra kernel may be rewritten as

$$K_m(s_1, \ldots, s_m) = \sum_{i_1=0}^{\infty} \cdots \sum_{i_m=0}^{\infty} C_{m,i_1,\ldots,i_m} \phi_{m,i_1}(s_1) \cdots \phi_{m,i_m}(s_m),$$

where $C_{m,i_1,\ldots,i_m}$ are the Volterra-Laguerre coefficients.

One of the most recent approaches for the Volterra model identification for computational mechanics problem was proposed by [Khawar *et al.* (2012)]. This approach is methodologically close to the Volterra-Laguerre method. Volterra kernels identification problem is addressed via expansion of the Volterra kernels in terms of scale functions and multi-wavelet functions employing multi-resolution. The resulting system is solved based on the least square method employing singular value decomposition. The method is applied for computational structural dynamics simulation of an aeroelastic wing in the paper by [Khawar *et al.* (2012)]. As footnote it is to be noted that in the paper by [Franz and Schölkopf (2006)] it is outlined that all the properties of discrete Volterra integral models theory are preserved by using polynomial kernels in a regularized regression framework.

11.2.1 *Kernels Identification via the Special Volterra Equations*

The objective of this section is to demonstrate applicability of numerical methods for the identification of transfer functions (Volterra kernels) from the special Volterra integral equations of the first kind we obtained using the training signals (11.2.1).

Before proceeding to the description of the numerical methods for the identification of Volterra kernels, let us concentrate on linear Volterra integral equations of the first kind which we write in the following form

$$\int_0^t x(t, s) K(s) ds = f(t), \quad 0 \le t \le T. \tag{11.2.9}$$

Here $x(t, s)$ is a known function of two variables defined in $\Delta = \{t, s/0 \le s \le t \le T\}$, $f(t)$ is a known function and the kernel $K(t)$ is the desired soution. Let $x_t' \in C_\Delta$, $f' \in C_{[0,t]}$, $f(0) = 0$ and, we assume $x(t, t) \ne 0$.

For a numerical solution of equation (11.2.9) one may employ classical methods of computational mathematics and methods for the solution of ill-posed problems solution.

In fact, Volterra equations of the first kind are a special case of Fredholm integral equations of the first kind known to be ill-posed in the Hadamard sense in any function spaces. Therefore, for the numerical solution of Volterra integral equations of the first kind one can use methods of regularization of ill-posed problems. Here readers may refer to the classical book of [Tikhonov and Arsenin (1977)]. For introduction to identification problems in terms of functional analysis readers may refer to the book of [Lorenzi (2001)] which focuses on an aspect of the theory of inverse problems, which is usually referred to as identification of parameters appearing in integrodifferential and differential equations. For classification of well-posed, ill-posed, and intermediate problems with practical examples readers my refer to the book of [Petrov and Sizikov (2005)]. Regularization of projection methods for solving ill-posed problems was considered by [Plato and Vainikko (1990)].

On the other hand, under the assumptions on initial data smoothness the problem (11.2.9) is well posed on $(\mathcal{C}_{[0,t]}, \overset{\odot\,(1)}{\mathcal{C}_{[0,t]}})$, where

$$\overset{\odot\,(1)}{\mathcal{C}_{[0,t]}} = \{f(t) : \ f(t) \in \overset{(1)}{\mathcal{C}_{[0,t]}}, \ f(0) = 0\}.$$

This fact allows the design of numerical solutions to equation (11.2.9) using the quadrature formulas.

For a detailed classification of the numerical methods for the Volterra equations solution the readers may refer to the book of [Brunner and Houwen (1986)]. It is to be noted that there are many numerical methods available for the numerical solution of the Volterra integral equations of the first kind (11.2.9) when the kernel of the equation $x(t, s)$ and the right-hand side $f(t)$ (inputs) are assumed to be known without error and thus have the desired smoothness.

If the input data are only known approximately $\tilde{x}(t, s), \tilde{f}(t)$, then there is a way that combines the positive aspects of both directions: error robustness and algorithmic simplicity.

The idea of this approach is the following. First, one can directly apply the discretization of the integral equations (11.2.9). But for noisy input data the selected method may not converge to the exact solution as $h \to 0$ Then one have to match the mesh step with the error level such that the discretization procedure itself has a regularizing property.

It is to be noted that trapezoid, midpoint, left and right rectangle quadrature rules enable a regularization algorithm where the step size itself is the regularization parameter.

Here the midpoint quadrature rule is used due to its algorithmic simplicity and due to to its ability to provide an approximate solution with an $\mathcal{O}(h^2)$ error in case of noise free input data. Below we consider the numerical solution to the Volterra equations for the identification of the kernels $K_1(t, s_1)$, $K_2(t, s_1, s_2)$. In both cases we prove second order convergence and demonstrate that the proposed method has the regularization property in sense of A. N. Tikhonov when one matches the step size with input data error level.

For didactical reasons let us first address the VIE of the 1st kind obtained above from the finite Volterra series model

$$\int_0^\omega K_1(t, s)\, ds = f_1(t, \omega), \ \ 0 \le s \le t \le T, \tag{11.2.10}$$

where $t, \omega \in \triangle_1 = \{t, \omega / 0 \le \omega \le t \le T\}$.

We introduce the uniform grid points as follows

$$t_j = jh, \ \omega_i = ih, \ \omega_{j-\frac{1}{2}} = (j - \frac{1}{2})h, \ i = \overline{1, N}, \ j = \overline{1, N}, \ Nh = T.$$

Application of the midpoint quadrature rule gives us the following system of linear algebraic equations of order $\frac{N(N+1)}{2}$

$$h \sum_{k=1}^i K^h_{1_{j,k-\frac{1}{2}}} = f^h_{1_{j,i}}, \ \ 1 \le i \le j \le N.$$

The uniqueness follows from the fact that the matrix of such SLAE is lower triangular with ones in the diagonal. Its solution is as follows

$$K^h_{1_{j,i-\frac{1}{2}}} = \frac{f^h_{1_{j,i}} - f^h_{1_{j,i-1}}}{h}, \ \ 1 \le i \le j \le N. \tag{11.2.11}$$

Let us investigate the convergence of K^h_1 to the exact K_1.

Theorem 11.2. *Let the solution of the integral equation (11.2.10) exists and let $f_1 \in C^{(3)}_{\triangle_1}$. Then midpoint quadrature method has second order of convergence and we have the following error estimate*

$$\|\varepsilon^h_1\|_{C_h} = \max_{i,j} \left| K_1(t_{jh}, \omega_{(i-\frac{1}{2})h}) - K^h_{1_{j,i-\frac{1}{2}}} \right|$$

$$\le \frac{1}{24} h^2 M_3,$$

$$\text{where } M_3 = \max_{t,\omega} \left| \frac{\partial^3 f_1}{\partial \omega^3} \right|.$$

Proof. Let us decompose the right hand side of the equation (11.2.11) into the Taylor series near the point $(t_j, \omega_{1_{i-\frac{1}{2}}})$, up to the 3rd order:

$$f_1(jh, ih) = f_1(jh,\ (i - \tfrac{1}{2})h) + \sum_{n=1}^{2} \frac{h^n}{n!} \left(\frac{1}{2} \frac{\partial}{\partial \omega} \right)^n f_1(jh, \omega) \Bigg|_{\omega=(i-\frac{1}{2})h}$$

$$+ \frac{h^3}{3!} \left(\frac{1}{2} \frac{\partial}{\partial \omega} \right)^3 f_1(jh, \omega) \Bigg|_{\omega=(i-\frac{1}{2})h+\theta_1 \frac{h}{2}} \ ,$$

$$f_1(jh, (i-1)h) = f_1(jh,\ (i - \tfrac{1}{2})h) + \sum_{n=1}^{2} \frac{h^n}{n!} \left(-\frac{1}{2} \frac{\partial}{\partial \omega} \right)^n f_1(jh, \omega) \Bigg|_{\omega=(i-\frac{1}{2})h}$$

$$+ \frac{h^3}{3!} \left(-\frac{1}{2} \frac{\partial}{\partial \omega} \right)^3 f_1(jh, \omega) \Bigg|_{\omega=(i-\frac{1}{2})h-\theta_2 \frac{h}{2}} \ ,$$

where $0 \le \theta \le 1$.

Substitution into (11.2.11) gives us

$$K^h_{1_{j,i-\frac{1}{2}}} = \frac{1}{h} \sum_{n=1}^{2} \frac{h^n}{n!2^n} \left(\left(\frac{\partial}{\partial \omega} \right)^n - \left(-\frac{\partial}{\partial \omega} \right)^n \right) f_1(jh, \omega) \Bigg|_{\omega=(i-\frac{1}{2})h}$$

$$- \frac{h^2}{3!} \left(-\frac{1}{2} \frac{\partial}{\partial \omega} \right)^3 f_1(jh, \omega) \Bigg|_{\omega=(i-\frac{1}{2})h-\theta_1 \frac{h}{2}}$$

$$- \frac{h^2}{3!} \left(\frac{1}{2} \frac{\partial}{\partial \omega} \right)^3 f_1(jh, \omega) \Bigg|_{\omega=(i-\frac{1}{2})h-\theta_2 \frac{h}{2}} \ .$$

Taking into account the exact solution

$$K_1(t, \omega) = f'_{1_\omega}(t, \omega),$$

we find the following difference

$$K_1\left(jh,\left(i-\frac{1}{2}\right)h\right) - K^{\cdot h}_{1_{j,i-\frac{1}{2}}}$$

$$= f'_{1_\omega}(t,\omega)\Big|_{t=jh,\ \omega=(i-\frac{1}{2})h} - \left(\frac{1}{2}\left(\frac{\partial}{\partial\omega}+\frac{\partial}{\partial\omega}\right)f_1(t,\omega)\right)\Big|_{t=jh,\ \omega=(i-\frac{1}{2})h}$$

$$- \frac{h^2}{3!}\left(-\frac{1}{2}\frac{\partial}{\partial\omega}\right)^3 f_1(jh,\omega)\Big|_{\omega=(i-\frac{1}{2})h-\theta_1\frac{h}{2}}$$

$$+ \frac{h^2}{3!}\left(-\frac{1}{2}\frac{\partial}{\partial\omega}\right)^3 f_1(jh,\omega)\Big|_{\omega=(i-\frac{1}{2})h-\theta_2\frac{h}{2}}.$$

Evaluating the absolute value of the resulting difference, we obtain the required error estimate for quadrature rule

$$\|\varepsilon^h\|_{C_h} = \max_{i,j}\left|K_1(t_{jh},\omega_{(i-\frac{1}{2})h}) - K^{\cdot h}_{1_{j,i-\frac{1}{2}}}\right| \le \frac{1}{24}h^2 M_3,$$

$$M_3 = \max_{t,\omega}\left|\frac{\partial^3 f_1}{\partial\omega^3}\right|,$$

where f_1 is three times differentiable function. We can see that midpoint quadrature method has second-order convergence. $\qquad\square$

Let us now consider the problem of identification of the kernel $K_2(t,s_1,s_2)$ from the following VIE

$$\int_0^{\omega_1}\int_0^{\omega_1} K_2(t,s_1,s_2)\,ds_1ds_2 - 2\int_0^{\omega_1}\int_{\omega_1}^{\omega_1+\omega_2} K_2(t,s_1,s_2)\,ds_1ds_2$$

$$+ \int_{\omega_1}^{\omega_1+\omega_2}\int_{\omega_1}^{\omega_1+\omega_2} K_2(t,s_1,s_2)\,ds_1ds_2 = f_2(t,\omega_1,\omega_2), \quad (11.2.12)$$

where $t,\omega_1,\omega_2 \in \triangle_2 = \{0 \le \omega_1+\omega_2 \le t \le T, \omega_1,\omega_2 \ge 0\}$.
  We introduce uniform grid points as follows

$$t_i = ih,\ \omega_{1_j} = jh,\ \omega_{1_{j-\frac{1}{2}}} = (j-\frac{1}{2})h,\ \omega_{2_k} = kh,\ \omega_{2_{k-\frac{1}{2}}} = \left(k-\frac{1}{2}\right)h,$$

$$i,j = \overline{1,N},\ k = \overline{0,N-1},\ Nh = T.$$

Application of the midpoint quadrature rule for the VIE (11.2.12) gives us the following system

$$h^2\left[\sum_{l=1}^{j}\sum_{m=1}^{j}K_{i,l-\frac{1}{2},m-\frac{1}{2}}^{h}-2\sum_{l=1}^{j}\sum_{m=j+1}^{j+k}K_{i,l-\frac{1}{2},m-\frac{1}{2}}^{h}+\sum_{l=j+1}^{j+k}\sum_{m=j+1}^{j+k}K_{i,l-\frac{1}{2},m-\frac{1}{2}}^{h}\right]$$

$$= f_{i,j,k}, \quad 0 \le j+k \le i \le N, \tag{11.2.13}$$

The number of equations is equal to the number of unknowns and its equal to

$$\sum_{n=1}^{N}\frac{n(n+1)}{2}=C_{N+2}^{3}. \tag{11.2.14}$$

This system has the following matix

$$\begin{pmatrix}
1 & 0 & 0 & 0 & 0 & 0 & 0 & 0 & 0 & 0 & \cdot \\
1 & 1 & -2 & 0 & 0 & 0 & 0 & 0 & 0 & 0 & \cdot \\
1 & 1 & 2 & 0 & 0 & 0 & 0 & 0 & 0 & 0 & \cdot \\
1 & 1 & -2 & 1 & 2 & -2 & 0 & 0 & 0 & 0 & \cdot \\
1 & 1 & 2 & 1 & -2 & -2 & 0 & 0 & 0 & 0 & \cdot \\
1 & 1 & 2 & 1 & 2 & 2 & 0 & 0 & 0 & 0 & \cdot \\
1 & 1 & -2 & 1 & 2 & -2 & 1 & 2 & 2 & -2 & \cdot \\
1 & 1 & 2 & 1 & -2 & -2 & 1 & 2 & -2 & -2 & \cdot \\
1 & 1 & 2 & 1 & 2 & 2 & 1 & -2 & -2 & -2 & \cdot \\
1 & 1 & 2 & 1 & 2 & 2 & 1 & 2 & 2 & 2 & \cdot \\
\cdot & \cdot & \cdot & \cdot & \cdot & \cdot & \cdot & \cdot & \cdot & \cdot & \cdot
\end{pmatrix}.$$

Diagonalization of the system (11.2.13) gives us an explicit inversion formulae

$$K_{11_{i,j+k-\frac{1}{2},j-\frac{1}{2}}}^{h}=\frac{f_{i,j-1,k+1}^{h}-f_{i,j-1,k}^{h}-f_{i,j,k}^{h}+f_{i,j,k-1}^{h}}{4\,h^2}, \tag{11.2.15}$$

$$1 \le j+k \le i \le N;$$

$$K_{11_{i,j-\frac{1}{2},j-\frac{1}{2}}}^{h}=\frac{f_{i,j-1,1}^{h}-2f_{i,j-1,0}^{h}+f_{i,j,0}^{h}}{2\,h^2}, \quad 1 \le j \le i \le N. \tag{11.2.16}$$

Let us investigate the convergence of K_2^h to exact solution K_2. The following Theorem can be proved similarly with the Theorem 11.2.

Theorem 11.3. *Let us suppose that the solution of the integral equation (11.2.12) exists and moreover $f_2 \in C_{\Delta_2}^{(4)}$. Then the midpoint quadrature* *method has second order of convergence and we have the following error estimate*

$$\|\varepsilon_1^h\|_{C_h} = \max_{i,j,k} \left| K_2\left(ih, \left(j+k-\frac{1}{2}\right)h, \left(j-\frac{1}{2}\right)h\right) - K_{2_{i,j+k-\frac{1}{2},j-\frac{1}{2}}}^h \right| \leq \frac{5}{48}h^2 M_4,$$

$$\text{where } M_4 = \max_{\substack{\mu+\nu=4 \\ t,\omega_1,\omega_2}} \left| \frac{\partial^4 f_2}{\partial \omega_1^\nu \partial \omega_2^\mu} \right|.$$

A Note on the Regularization The idea to employ the mesh step size as a regularization parameter was initially proposed for the weakly singular Fredholm equations in [Dmitriev and Zaharova (1968)].

We assume that instead of the exact source part of the equation $f_2(t, \omega_1, \omega_2)$ we have an approximate one $\widetilde{f}_2(t, \omega_1, \omega_2)$ with the following estimate

$$\left| f_2(t, \omega_1, \omega_2) - \widetilde{f}_2(t, \omega_1, \omega_2) \right| \leq \delta. \tag{11.2.17}$$

Based on (11.2.15) and (11.2.16), by using the triangle inequality we have the following error estimate

$$\widetilde{\varepsilon}_h = \max_{i,j,k} \left| K_2 - \widetilde{K}_2^h \right| \leq \frac{2\delta}{h^2} + \frac{5}{48}h^2 M_4, \tag{11.2.18}$$

where $\widetilde{K}^h$ denote the solution to the system (11.2.13) with $\widetilde{f}$ as the source part of the equation.

This estimate shows the cumulative influence of two errors: due to the input data (the first term) and due to the method (the second term). Let us define the quasi-optimal step as follows

$$h_{\text{qo}} = \arg \min_h \left(\frac{2\delta}{h^2} + \frac{5}{48}h^2 M_4 \right).$$

Therefore we get the following estimates:
for the step

$$h_{\text{qo}}(\delta) \asymp \delta^{\frac{1}{4}}, \tag{11.2.19}$$

for the error

$$\varepsilon_{h_o}(\delta) = \mathcal{O}(\delta^{\frac{1}{2}}). \tag{11.2.20}$$

These estimates show the regularization property for the cubature method. It means that error of the numeric solution tends to zero at a rate not less than $\delta^{1/2}$ as $\delta \to 0$ and step selection of the order $\delta^{1/4}$.

Table 11.1

h	ε^h
1.00000	0.16666
0.50000	0.04166
0.25000	0.01041
0.12500	0.00260
0.06250	0.00065
0.03125	0.00016
0.01562	0.00004
0.00781	0.00001

Numeric Examples Let us now consider an illustrative numerical examples. First we consider (11.2.12) for the kernel $K_2(t, s_1, s_2) = s_1^2 + s_2^2$:

$$\int_0^{\omega_1} \int_0^{\omega_1} (s_1^2 + s_2^2) \, ds_1 ds_2 - 2 \int_0^{\omega_1} \int_{\omega_1}^{\omega_1+\omega_2} (s_1^2 + s_2^2) \, ds_1 ds_2$$

$$+ \int_{\omega_1}^{\omega_1+\omega_2} \int_{\omega_1}^{\omega_1+\omega_2} (s_1^2 + s_2^2) \, ds_1 ds_2 = f(t, \omega_1, \omega_2),$$

then

$$f_2(t, \omega_1, \omega_2) = -\frac{4}{3}\omega_1\omega_2^3 - 4\omega_1^2\omega_2^2 - \frac{16}{3}\omega_1^3\omega_2 + \frac{2}{3}(\omega_1 + \omega_2)^4. \qquad (11.2.21)$$

Differentiation of f gives us

$$K_2(t, \omega_1, \omega_1 + \omega_2) = 2\omega_1^2 + 2\omega_1\omega_2 + \omega_2^2. \qquad (11.2.22)$$

Direct substitution ω_1 and $\omega_1 + \omega_2$ instead of s_1, s_2 into the kernel $K_2(t, s_1, s_2) = s_1^2 + s_2^2$ gives $K_2(t, \omega_1, \omega_1 + \omega_2) = 2\omega_1^2 + 2\omega_1\omega_2 + \omega_2^2$.

For specific example with kernel $K_2(t, s_1, s_2) = s_1^2 + s_2^2$ the midpoint rule was employed and an error values are obtained for various step sizes. One can see the second order convergence from the Table 11.1. Asymptotic estimates for h_{qo} and for $\varepsilon_{h_{qo}}$ were approved with these examples. For various kernels the optimal step has been computed for the perturbed source part of the equation $\tilde{f}_\delta$. The sawtooth pattern perturbations have been selected. The Fibonacci method (with 10 tests) has been employed for the optimal step computation. Here readers may refer here for example to the book of [Wilde (1964)]. The following notations are used: δ is input data error, h_{op} is optimal step, ε_{op}^h is optimal error, ℓ is the current interval width.

One can see from Table 11.2 and Table 11.3 that when the input data errors δ is reduced by 10^4 times, h_{op} reduces by 10 times and ε_{op}^h by 100 times which is in line with asymptotic estimates (11.2.19), (11.2.20).

Table 11.2 Results for the case of
$K_{11}(t, \nu_1, \nu_2) = \nu_1^2 + \nu_2^2$

δ	h_{op}	$\varepsilon_{\mathrm{op}}^h$	ℓ
10^{-1}	0.9887	0.36751	0.0112
10^{-2}	0.5888	0.11547	0.0111
10^{-3}	0.3307	0.03651	0.0066
10^{-4}	0.1858	0.01154	0.0037
10^{-5}	0.1044	0.00365	0.0020
10^{-6}	0.0586	0.00115	0.0011
10^{-7}	0.0329	0.00036	0.0006

Table 11.3 Results for the case of
$K_{11}(t, \nu_1, \nu_2) = e^{\nu_1} e^{\nu_2}$.

δ	h_{o}	$\varepsilon_{\mathrm{o}}^h$	ℓ
10^{-1}	0.64044	0.72427	0.0112
10^{-2}	0.43896	0.16410	0.0071
10^{-3}	0.22687	0.05988	0.0049
10^{-4}	0.15550	0.01942	0.0025
10^{-5}	0.08037	0.00651	0.0017
10^{-6}	0.04515	0.00216	0.0009

Table 11.4 Results for the case of
$K_{11}(t, \nu_1, \nu_2, \nu_3) = \nu_1 \nu_2$.

δ	h_{o}	$\varepsilon_{\mathrm{o}}^h$	ℓ
10^{-1}	0.9887	0.20457	0.0112
10^{-2}	0.9776	0.02092	0.0111
10^{-3}	0.9666	0.00214	0.0109
10^{-4}	0.9558	0.00021	0.0108
10^{-5}	0.9450	0.00002	0.0107
10^{-6}	0.9344	0.00000	0.0106
10^{-7}	0.9239	0.00000	0.0104

Table 11.4 shows the special kernel case such as method error $\frac{5}{48}h^2 M_4$ is zero and $\arg\min_{h} \frac{2\delta}{h^2}$ is achieved as $h_{\mathrm{op}} \to \max_{t\in[0,1]} t$.

11.3   Impulse Control as Solution of Multiple Integral Equation

Let us now address the following equation, where transfer functions K_m are convolution kernels and they are supposed to be known

$$\sum_{m=1}^{N} \int_0^t \cdots \int_0^t K_m\left(t - s_1, \ldots, t - s_m\right) \prod_{i=1}^{m} x(s_i) ds_i$$

$$- f\left(t\right) = 0, \; 0 \le t \le T. \tag{11.3.1}$$

In feed-back control flow process based on the Volterra models (see, e.g., article of [Belbas and Bulka (2010)]) one needs to find input signal $x(t)$ taking into account known, or desired $f(t)$, i.e., to solve the nonlinear VIE of the 1st kind w.r.t input signal $x(t)$.

Based on the assumption that the transfer functions are known and the Volterra model is already constructed, and the objective response $f(t)$ of the dynamic system is defined it is natural to make the following problem statement. In order to find the input signal $x(t)$ (control), which is the solution of the polynomial VIE of the first kind (11.3.1). [Apartsyn (2003)], [Belbas and Bulka (2010)] and [Sidorov and Sidorov (2012)] derived the results concerning the continuous solution of equation (11.3.1). It is to be noted that only for $f(0) = 0$ the equation (11.3.1) may have the classical continuous solution.

Another principal challenge in this theory is that continuous solution may exist only locally on the certain interval. Outside of this interval the solution may blow-up or can start branching. So it appears to be natural to search for the solution in the class of generalized functions and sum of regular component and the singular component containing the combination of the Dirac delta functionals.

Impulse control is an important topic in the theory of nonlinear and linear dynamic systems, here readers may refer to the books of [Zavalishin and Sesekin (1997)], [Tsypkin and Popkov (1973)] and [Miller and Rabinovich (2003)]. In this chapter we formulate the existence theorem of generalized solution of the equation (11.3.1). We would also demonstrate that number of generalized solutions is equal to the number of roots of the certain polynomial. The solutions consist of singular part with point-wise support (here readers may refer to the textbooks of [Vladimirov (1979)], [Kanwal (2013)]) and the regular part, which satisfies the special VIE. We employ the results from Chapter 9 and demonstrate that equation (11.3.1) can have several generalized solutions and the construction algorithm is proposed.

Let us assume that Volterra kernels in our model (11.3.1) be separable, following the Definition 11.1. Therefore they can be presented as follows

$$K_m (t - s_1, \ldots, t - s_m) = Q_{m,1} (t - s_1)$$
$$\times \cdots \times Q_{m,m} (t - s_m), \quad m = \overline{1, N}.$$

For sake of clarity we will suppose below the complete symmetry of the kernels:

$$K_m (t - s_1, \ldots, t - s_m) = \prod_{i=1}^{m} Q_m (t - s_i). \qquad (11.3.2)$$

We take into account equality (11.3.2) and rewrite the equation (11.3.1) as follows

$$\sum_{m=1}^{N} \left(\int_0^t Q_m (t - s) x (s) \, ds \right)^m = f(t), \quad 0 \leq t \leq T. \qquad (11.3.3)$$

Based on the terminology proposed by [Belbas and Bulka (2010)], equation (11.3.3) is so-called *multiple* integral equation of the first kind. Based on the methods proposed in the papers of [Sidorov and Sidorov (2006, 2011a)] we introduce the following definitions

Definition 11.3. The set of indefinitely differentiable finite functions $s(t)$ with supports on the interval $(0, T)$ we denote by $D_{(0,T)}$. The set of linear continuous functionals

$$x \in \mathcal{L}(D_{(0,T)} \to \mathbb{R}^1)$$

we call *generalized functions space* and denote by $D'_{(0,T)}$.

Definition 11.4. x from $D'_{(0,T)}$ we call *generalized solution* of equation $F(x, t) = 0$, if $\forall s(t) \in D_{(0,T)}$ the equality

$$\int_{-\infty}^{+\infty} F(x, t) \, s(t) \, dt = 0$$

is satisfied.

Let us now follow [Sidorov and Sidorov (2006)] and search for the solution of equation (11.3.3) in class $D'_{(0,T)}$ as follows

$$x(t) = c\delta(t) + v(t, c). \tag{11.3.4}$$

Here $\delta(t)$ denotes the Dirac delta functional, constant c and continuous function $v(t, c)$ are assumed to be unknown.

Theorem 11.4. *Let c be the simple root of the polynomial $L(c) - f(0)$, here*

$$L(c) = \sum_{m=1}^{N} (Q_m(0)c)^m,$$

and let all the kernels and function $f(t)$ be differentiable for $0 \le t \le T$. Then equation (11.3.3) possesses the solution $x(t) = c\delta(t) + v(t, c)$ in class $D'_{(0,T)}$, where continuous function $v(t, c)$ depends on choice of root c and it is solved uniquely using the successive approximations method or other techniques.

If polynomial $L(c) - f(0)$ has multiple root then corresponding solution of equation (11.3.3) may contain derivative of the Dirac $\delta(t)$ delta functional. This case can be studied in a similar way.

Since equation (11.3.3) is nonlinear then regular part v of generalized solution (11.3.4) is defined in the small neighborhood $[0, \rho^*)$, where $\rho^* \le T$. Regular part of solution $v(t, c)$ is continuous function which exists on $[0, \rho^*)$. For continuation of regular part $v(t)$ in to the region $t > \rho^*$ even in analytical case can be found a point $t^* \in (\rho, T]$, for which

$$\lim_{t \to t^* - 0} v(t, c^*) = \infty.$$

In this point the *"blow-up"* phenomena occurs which corresponds to loosing of system control. The lower boundary of the interval of existence of the function $v(t)$ can be defined using the monotone majorants technique. For more details, refer to the paper [Sidorov and Sidorov (2012)] and illustrative example below.

Let us note that if $f(0) = 0$, $Q_1(0) \ne 0$, then $c = 0$ is simple of $L(c)$ and the corresponding solution of equation (11.3.3) is classical one. If also the polynomial $L(c)$ has simple non zero roots then in addition to the classical one there exist generalized solutions.

Above it was supposed that not all the $Q_n(0) = 0$. Now let us address the case when

$$Q_n(0) = Q'_n(0) = \cdots = Q_n^{(k-1)}(0) = 0, n = 1, 2, \ldots, N,$$

$$\sum_{n=1}^{N} |Q_n^{(k)}(0)| \neq 0.$$

Let us introduce the *characteristic equation*

$$L(c) = \sum_{n=1}^{N} (Q_n^{(k)}(0)c)^n = f(0).$$

Let c^* be the simple root, i.e., $L'(c^*) \neq 0$ then the integral equation (11.3.2) will have the next solution

$$x(t) = c^* \delta^{(k)} + c_{k-1} \delta^{(k-1)} + \cdots + c_0 \delta(t) + v(t).$$

Here constants $c_{k-1}, \ldots, c_0$ can be uniquely defined from the recurrence sequence of the linear algebraic equations. Continuous function $v(t)$ can be defined in the neighborhood of the point $t = 0$ from the regular VIE of the second kind.

Remark 11.1. In the general case when the transfer functions are representable as follows

$$K_m(t, s_1, \ldots, s_m) = Q_{m,1}(t - s_1)$$

$$\times \ldots \times Q_{m,m}(t - s_m), \quad m = \overline{1, N}$$

the above mentioned theorem remains correct, but polynomial L will be as follows

$$L(c) = \sum_{m=1}^{N} \prod_{i=1}^{m} Q_{m,i}(0) c^m.$$

11.4 Illustrative Examples

If $f(t)$ and $K_m(t)$ are analytical functions, then regular part of the solution can be found as series

$$v = \sum_{i=0}^{\infty} v_i t^i. \tag{11.4.1}$$

Its coefficients can be computed with method of undetermined coefficients. It is easy to estimate below the radius of its convergence using the method of convex majorants [Khromov (2006)].

Example 11.1. Let us consider the equation

$$w(t) = \left(\int_0^t t s^2 w(s) ds \right)^2 + t, \quad t \geq 0.$$

Taking into account [Sidorov and Sidorov (2012)] the majorant system

$$\begin{cases} r = r^2 \frac{\rho^4}{3} + \rho \\ 1 = \frac{2}{3} r \rho^4 \end{cases}$$

has the solution $\rho^* = \left(\frac{3}{4}\right)^{1/5}, r^* = 2\left(\frac{3}{4}\right)^{1/5}$. Hence this integral equation for $|t| \leq \rho^*$ possesses an analytical solution.

Example 11.2. For equation

$$\int_0^t x(s)\,\mathrm{d}s + \left(\int_0^t x(s)\,\mathrm{d}s\right)^2 = 2 + 2t - t^2, \quad t \geq 0$$

the conditions of the theorem are satisfied. The corresponding polynomial $c^2 + c - 2 = 0$ has the simple roots $1, -2$, which correspond to two generalized solutions

$$x_1(t) = \delta(t) + \frac{2}{3} + \frac{10}{27}t + \mathcal{O}(t^2),$$

$$x_2(t) = -2\delta(t) - \frac{2}{3} - \frac{10}{27}t + \mathcal{O}(t^2).$$

Example 11.3. Let us consider the equation

$$\left(\int_0^t x(s)ds\right)^2 + \int_0^t x(s)ds + t = 0, \ t > 0.$$

The corresponding characteristic equation $c^2 + c = 0$ has two simple roots $c_1 = 0$, $c_2 = -1$. Thus it has two solutions $x_1(t) = -\delta(t) - \frac{1}{\sqrt{1-4t}}$, $x_2(t) = \frac{1}{\sqrt{1-4t}}$, $t_0 = \frac{1}{4}$ is blow-up point of these solutions.

As the footnote let us notice that the solutions addressed here are the generalized (see the definition above, refer [Shwartz (1961)] for more details) and the Dirac delta functional in applications can be (as example) considered as limit (in the sense of distributions) of the sequence $\left\{\frac{1}{\pi}\frac{\sin(\omega t)}{t}\right\}$ for $\omega \to \infty$ or as sequence of Gaussians $\delta_a(t) = \lim_{\sigma \to 0} \frac{1}{\sigma\sqrt{\pi}}e^{\frac{-t^2}{\sigma^2}}$.

Remark 11.2. The main classical solutions in sense of L.V. Kantorovich [Kantorovich *et al.* (1950)] of nonlinear VIE are studied in Chapter 9, where the more general case is addressed. Convergence of the successive approximations is established through studies of majorant integral and majorant algebraic equations. For numerical solution in the neighborhood of such points it make sense to employ an adaptive meshes.

We presented the theoretical studies on the Volterra nonlinear systems theory. The nonlinear (polynomial) VIE of the first kind describing the control process of nonlinear dynamic processes of "input-output" type may have the generalized solutions. Generalized solutions of these equations can be constructed as the sum of the singular component with point support and a regular function. The coefficients of the singular part have been determined from the nonlinear algebraic equations. Regular function is uniquely determined by successive approximations.

In the design of the control of nonlinear dynamic systems with memory on the basis of the functional Volterra series it is necessary to take into account that solution is not unique and some optimal solution criteria should be selected.

In Section 12.1, we present the application of the Volterra series for modelling of nonlinear dynamics of heat transfer in a channel with single phase coolant.

PART 3
Integral Models Applications

Part Summary

The objective of this part is to demonstrate the applications of the integral models in three different fields: heat and power engineering, electric power engineering and in multidimensional digital signal processing.

In Section 12.1 the finite Volterra series are employed to model the non-linear dynamics of heat exchanger. In Section 12.2 the Volterra models with piecewise continuous kernels are discussed in framwork of modeling evolving dynamical energy systems. Section 13 considers non-stationary patterns removal which are result of a multiplicative superposition rule. Such patterns appear in digital image and motion pictures restoration problem. Finally in Section 13.6 two integral models applications in electric power engineering problems are addressed. First, problem of electric power systems parameters forecasting is addressed. Then the method for robust detection of inter-area electro-mechanical unstable oscillations in power systems using dynamic data is proposed and tested on real and synthetic data.

Chapter 12

The Volterra Models Applications

12.1 Identification of Finite Volterra Series Models for Heat Exchange Processes

12.1.1 *Introduction*

The mathematical modelling and numerical simulation of dynamic behavior of thermal power plants particularly the steam generators and heat exchangers, important for a wide range of tasks, covering nearly all the stages starting from the initial design of the thermal power plants to their operation. Heat exchange is a process that is used to change the temperature distribution of two materials, when they are in direct or indirect contacts.

The reference model (here readers may refer to the paper of [Tairov (1989)]) of nonlinear dynamics of heat transfer in a channel with single phase coolant is used in order to generate the synthetic data for construction and testing the Volterra integral models of nonlinear dynamics. A model of a heat exchange process that occurs in the components of a heat exchanger with independent heat supply was used as a reference. According to [Tairov (1989)] the deviation of the enthalpy at the output $\Delta i\left(t\right)$ at arbitrary changes of liquid flow rate disturbances $\Delta D\left(t\right)$ and heat supply $\Delta Q(t)$ is defined by the following relationship:

$$\Delta i\left(t\right) = \frac{\lambda_1 \lambda_2}{\lambda_2 - \lambda_1} \int_0^t \left(\Delta Q(s) - \frac{Q_0}{D_0} \Delta D(s) \right)$$

$$\times \left(e^{-\lambda_1 \int_s^t D(s_1)ds_1} - e^{-\lambda_2 \int_s^t D(s_1)ds_1} \right) ds. \qquad (12.1.1)$$

In the relationship (12.1.1) t indicates the time, λ_1 and λ_2 are known constant values, the parameters with index '0' is used to denote parameters' initial stationary condition, Δ is an increment to a corresponding parameter of the initial stationary condition, e.g., $D(t) = D_0 + \Delta D(t)$. This reference model has been employed to generate the systems' reaction output signals on the above introduced signals $x_{\omega_1,...,\omega_{m-1}}^{\alpha_k}(t)$ for the Volterra model identification.

The Volterra non-stationary model is presented in paper of [Sidorov (2002)]. The results of experiments with product integration methods and "optimal" amplitudes selection are shown in papers of [Apartsyn *et al.* (2013); Apartsyn and Solodusha (2004)].

For more information regarding the experience of using the Volterra models for identification and control of heat exchange processes using the synthetic reference data readers may refer to [Apartsyn *et al.* (2013)], some details regarding the practical employment of the Volterra models for power consumption optimization for a gas turbine compressor are presented in the paper of [Scherbinin (2010)].

12.1.2 *The Heat Transfer in a Channel with Single Phase Coolant*

Partial differential equations are the conventional tools to model the dynamical processes in heat power plants which enable the creation of mathematical models of the process under study.

[Tairov and Zapov (1991)] have demonstrated that the following algebraic-differential system

$$\Delta D(\tau)\frac{di_0^*}{dz} + D(\tau)\frac{\partial \Delta i^*(\Delta p, \Delta t)}{\partial z} + g_m\frac{\partial \Delta i^*(\Delta p, \Delta t)}{\partial \tau}$$
$$= \Delta\alpha(\tau)h(\theta_0 - t_0) + \alpha(\tau)h(\Delta\theta(z,\tau) - \Delta t(z,\tau)), \qquad (12.1.2)$$

$$\Delta q(\tau) - g_m c_m\frac{\partial \Delta\theta(z,\tau)}{\partial \tau}$$
$$= \Delta\alpha(\tau)h(\theta_0 - t_0) + \alpha(\tau)h(\Delta\theta(z,\tau) - \Delta t(z,\tau)) \qquad (12.1.3)$$

describes dynamical heat-exchange process.

Here τ is the time (sec), z is the axis coordinate (m), D is the flow of matter (kg/sec); q is the heat load (kWh), g is the weight (kg/m), h is the heat transfer surface (length) (m); i^* is the enthalpy (kJ/kg), t and θ are the temperatures of the flow and of the wall (K) respectively, c is the specific heat (kJ/kg $\cdot$ K), p is the pressure (N/m^2), α is the heat transfer coefficient (kWh/m$^2 \cdot$ K), Δ is an increment, e.g., $D(\tau) = D_0 + \Delta D(\tau)$, $\alpha(\tau) = \alpha_0 +$

$\Delta\alpha(\tau)$; index 0 stands for initial value, e.g., $i_0^* = i^*(\Delta p, \Delta t)|_{\tau=0}$, "in" is value at input instant, "fm" is flow of matter, and finally "m" stands for the wall material.

The initial conditions are as follows

$$\Delta i^*(\Delta p, \Delta t)|_{\tau=0} = 0, \quad \Delta t(z, \tau)|_{\tau=0} = 0, \quad \Delta\theta(z, \tau)|_{\tau=0} = 0.$$

Boundary conditions are as follows

$$\Delta i^*(\Delta p, \Delta t)|_{z=0} = \Delta i_{in}(\tau).$$

The system parameters Δi^*, Δt are linked with each other by the following equation

$$\Delta i^*(\Delta p, \Delta t) = c_m \Delta t(z, \tau) + \left(\frac{\partial i^*}{\partial p}\right)_t \Delta p(\tau). \tag{12.1.4}$$

For arbitrary function $D(\tau)$ it seems to be impossible to find an analytical solution to the problems (12.1.2)–(12.1.4) for arbitrary $D(\tau)$.

Let us make the notation $i^*(\Delta p, \Delta t) = i(\tau)$ assuming a linear change in the spatial variable z. We proceed from a model with distributed parameters (12.1.2), (12.1.3) to a model with lumped parameters

$$\Delta D(\tau)(i_0 - i_{0in}) + D(\tau)(\Delta i(\tau) - \Delta i_{in}(\tau)) + G\frac{d\Delta i(\tau)}{d\tau}$$
$$= \Delta\alpha(\tau)H(\theta_0 - t_0) + \alpha(\tau)H(\Delta\theta(\tau) - \Delta t(\tau)), \tag{12.1.5}$$

$$\Delta Q(\tau) - G_m c_m \frac{d\Delta\theta(\tau)}{d\tau}$$
$$= \Delta\alpha(\tau)H(\theta_0 - t_0) + \alpha(\tau)H(\Delta\theta(\tau) - \Delta t(\tau)), \tag{12.1.6}$$

with following initial conditions $\Delta i(\tau)|_{\tau=0} = 0$, $\Delta t(\tau)|_{\tau=0} = 0$, $\Delta\theta(\tau)|_{\tau=0} = 0$.

Here $Q = q \cdot l$ is the total heat load (kWh), $G = g \cdot l$ is the total weight (kg), $H = h \cdot l$ is the total surface of the heat-exchange (m^2), l is the length of the sector under study.

[Tairov (1989)] assumed $\alpha = K_\alpha D$, where K_α is the constant value and the solution to the problems (12.1.4)–(12.1.6) is obtained. Note that at the initial time we have

$$i_0 - i_{0in} = \frac{\alpha_0}{D_0}H(\theta_0 - t_0) = K_\alpha H(\theta_0 - t_0) = \frac{Q_0}{D_0}.$$

Then the system (12.1.5) and (12.1.6) can be presented as follows

$$\frac{d\Delta i(\tau)}{d\tau} + (a_1\Delta i(\tau) + b_1\Delta\theta(\tau))D(\tau) = g(\tau), \tag{12.1.7}$$

$$\frac{d\Delta\theta(\tau)}{d\tau} + (a_2\Delta i(\tau) + b_2\Delta\theta(\tau))D(\tau) = \omega(\tau), \qquad (12.1.8)$$

where $a_1 = \dfrac{1}{G_{\text{fm}}}\left(1+\dfrac{K_\alpha H}{c_{\text{fm}}}\right), a_2 = -\dfrac{K_\alpha H}{G_{\text{fm}}c_{\text{fm}}c_{\text{m}}}, b_1 = -\dfrac{K_\alpha H}{G_{\text{m}}}, b_2 = \dfrac{K_\alpha H}{G_{\text{m}}c_{\text{m}}},$

$$g(\tau) = \frac{D(\tau)}{G_{\text{fm}}}\left(\Delta i_{\text{in}}(\tau) - K_\alpha H K_p \Delta p(\tau)\right), K_p = -\frac{1}{c_{\text{fm}}}\left(\frac{\partial i}{\partial p}\right)_t,$$

$$\omega(\tau) = \frac{1}{G_{\text{m}}c_{\text{m}}}\Delta Q(\tau) - \frac{Q_0}{D_0 G_{\text{m}}c_{\text{m}}}\Delta D(\tau) + \frac{K_\alpha H K_p}{G_{\text{m}}c_{\text{m}}}D(\tau)\Delta p(\tau).$$

The system (12.1.7) and (12.1.8) can be solved by reduction to the following differential equations with variable coefficients

$$\frac{dR_{1,2}(\tau)}{d\tau} + s_{1,2}D(\tau)R_{1,2}(\tau) = \alpha_{1,2}g(\tau) + \beta_{1,2}\omega(\tau),$$

where $R_{1,2}(\tau) = \alpha_{1,2}\Delta i(\tau) + \beta_{1,2}\Delta\theta(\tau)$ $s_{1,2}$ are roots of the characteristic equation of the system (12.1.7)–(12.1.8); $\alpha_{1,2}, \beta_{1,2}$ are the constant values obtained under the following assumption $\alpha_{1,2}a_1 + \beta_{1,2}a_2 = s_{1,2}\alpha_{1,2}, \alpha_{1,2}b_1 + \beta_{1,2}b_2 = s_{1,2}\beta_{1,2}$.

Let us assume below $\Delta i_{\text{in}}(\tau) = 0$, $\Delta p(\tau) = 0$, and $\Delta Q(\tau) = 0$, and the output enthalpy deviation $\Delta i(t)$ for arbitrary laws of perturbation flow $\Delta D(\tau)$ and under the stationary value of the heat Q_0 can be presented (here readers may refer to [Tairov and Zapov (1991)]) as follows

$$\Delta i(\tau) = \frac{b_1(R_1(\tau) - R_2(\tau))}{s_1 - s_2}, \qquad (12.1.9)$$

where

$$R_{1,2}(\tau) = e^{-s_{1,2}\int_0^\tau D(\eta)d\eta}\int_0^\tau\left(\frac{\Delta Q(\eta)}{G_{\text{m}}c_{\text{m}}} - \frac{Q_0\Delta D(\eta)}{D_0 G_{\text{m}}c_{\text{m}}}\right)e^{s_{1,2}\int_0^\eta D(\xi)d\xi}d\eta.$$

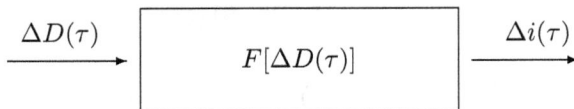

$$\Delta D(\tau) \longrightarrow \boxed{\quad F[\Delta D(\tau)] \quad} \longrightarrow \Delta i(\tau)$$

Fig. 12.1.2.

It is to be noted that $\Delta i(\tau)$ is strongly nonlinearly influenced with $\Delta D(\tau)$.

To analyze and evaluate the accuracy of modelling we will use the finite Volterra models (linear, quadratic and cubic models), and we employ the reference model (12.1.9) to generate the synthetic data without using the experiments with real heat-exchanger.

The problem of mathematical modelling a dynamic systems based on the finite Volterra models can be divided into the following stages:

(1) Recording of a set of dynamic responses of the system to the special training (piecewise constant) signals is used;
(2) Identification of the Volterra kernels (transfer functions) and use them in actual Volterra model.

Let us write the cubic model we used in the experiments

$$\int_0^\tau K_1(s)\Delta D(\tau - s)\, ds + \int_0^\tau \int_0^\tau K_{11}(s_1, s_2)\Delta D(\tau - s_1)\Delta D(\tau - s_2)\, ds_1 ds_2$$

$$+ \int_0^\tau \int_0^\tau \int_0^\tau K_{111}(s_1, s_2, s_3)\Delta D(\tau - s_1)\Delta D(\tau - s_2)\Delta D(\tau - s_3) ds_1 ds_2 ds_3$$

$$= \Delta i(\tau), \tag{12.1.10}$$

where $\Delta i(\tau) = i(\tau) - i_0$, $i_0 = \frac{Q_0}{D_0} + i_{\text{in}}$. The Volterra kernels K_1, K_{11}, K_{111} have been identified for stationary values of $D_0 = 0,16$ kg/sec and $Q_0 = 100$ kWh using the above presented method. In our numerical experiments the following input signals have been used to identify the Volterra kernels

$$\Delta D_{\omega_1, \omega_2}(\tau) \stackrel{\text{def}}{=} x_{\omega_1, \omega_2}(\tau) = C \cdot (e(t) - 2e(t - \omega_1) + e(t - \omega_1 - \omega_2)),$$

where $0 \le \omega_1 \le \omega_1 + \omega_2 \le \tau \le T$, $C \in \{0.04, 0.08\}$, $T = 40$ is transient time. For algorithms regarding the optimal selection of the training input signals' amplitudes readers may refer to the paper of [Apartsyn and Solodusha (2004)].

The numerical experiments with the reference model (12.1.9) have demonstrated the diverse behavior of the linear, quadratic and cubic Volterra models, see Fig. 12.1.

Table 12.1 and Table 12.2 represents summary of the duration of system responses depending on the time of the transition process T and quantization step h for the quadratic and cubic models construction.

Table 12.1   Responses duration for quadratic model identification

formula	responses amount
$2\left(\frac{T}{h}+1\right)$	42

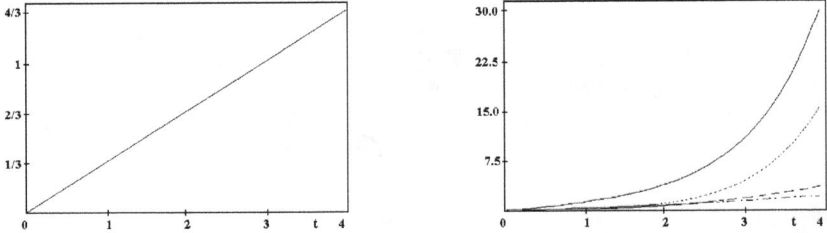

Fig. 12.1.   **Left:** input signal. **Right:** reactions of the Volterra models. Solid line marks reaction of the reference model, short dash line stands for cubic model signal, long dash line stands for quadratic model signal and pointed dash line marks the linear model reaction.

Table 12.2   Responses duration for cubic model identification

formula	responses amount
$6\frac{T}{h}\left(\frac{T}{h}+1\right)$	2520

As a footnote it is to be outlined that our numerical experiments have demonstrated the significance of the optimal amplitudes selection for the training input signals. In a number of test signals the cubic model was more accurate as shown in Fig. 12.1, but the accuracy of the quadratic model was more stable during the experiments with synthetic data.

12.2   Evolving Dynamical Systems Modelling

12.2.1   *Problem Statement*

In this section we consider the application of the VIE of the first kind with piecewise continuous kernels for evolving dynamical systems modelling.

Results described in present chapter are obtained in collaboration with E. Markova [Markova and Sidorov (2014)], A. Tynda and I. Muftahov [Sidorov *et al.* (2014)].

The equipment replacement process is conventionally modeled in discrete time in the operations research community. The optimal replacement strategies are found from discrete (or integer) programming problems, well known for their analytical and computational complexity. Here readers may refer to the publication of [Espiritu and Coit (2008)] for more details regarding replacement analysis methodology for the system-level component replacement schedules determination for electricity distribution systems composed of sets of heterogeneous assets. [Schwarz (2005)] has employed the mixed integer linear programming based optimization model to assess the economically motivated construction of new conventional power plants, and the modernization of existing conventional power plants in Germany. [Spiecker and Weber (2011)] assesses electricity market development in the presence of stochastic power feed-in and endogenous investments of power plants and renewable energies up to 2050 using linear stochastic programming model.

An alternative approach was represented by continuous-time Glushkov models and described by the Volterra integral equations of the first kind with jump discontinuous kernels which theory have not been fully studied to the best of our knowledge and fully addressed in the present book.

Let us briefly outline the two-sector Glushkov integral model of evolving dynamical systems (for more details readers may see the references in the publications of [Hritonenko and Yatsenko (1996, 1999); Apartsyn (2003); Karaulova and Markova (2008); Ivanov *et al.* (2006)]). Such integral models enable a qualitative study of the dynamical processes for replacing outdated equipment since they take into account age of the industrial capacities and the dynamics of the system development over the prehistory.

First we refer to the two-sector macroeconomic model from the publications of [Hritonenko and Yatsenko (1996, 1999); Apartsyn (2003)]. This model describes the interrelation of two groups in macroeconomics, producing means of production (subsystem A) and producing customer goods (subsystem B) and is given by the following model

$$x(t) = \int_{\alpha_1(t)}^{t} \alpha(t,s)x(s)y(s)\,ds, t \in [t_0, T], \tag{12.2.1}$$

$$p\left(t\right) = \int_{\alpha_2(t)}^{t} \beta(t,s)x\left(s\right)\left[1 - y\left(s\right)\right]ds, \qquad (12.2.2)$$

$$c(t) = \int_{\alpha_1(t)}^{t} y(s)x(s)ds + \int_{\alpha_2(t)}^{t} [1 - y(s)]x(s)\,ds, \qquad (12.2.3)$$

$$0 \leq y\left(t\right) \leq 1, 0 \leq \alpha_i\left(t\right) < t, t \in [t_0, T] \qquad (12.2.4)$$

$$N(t) = \lambda(t)x(t) + p(t), \qquad (12.2.5)$$

$$\alpha'_{1,2}\left(t\right) \geq 0, t \in [t_0, T], \qquad (12.2.6)$$

where:

$x\left(t\right)$ is electric power cumulative commissioning per time unit,

$y(t)$ is relative partition of electric power commissioning per time unit and those which remain in the subsystem A,

$\alpha\left(t,s\right)$ is efficiency in the subsystem A,

$\beta\left(t,s\right)$ is efficiency in the subsystem B,

$\alpha_1\left(t,s\right)$ is upper time limit of equipment dismounting in the subsystem A,

$\alpha_2\left(t,s\right)$ is upper time limit for equipment dismounting in the subsystem B,

$c(t)$ is cumulative amount of the elements exploited per time unit t,

$N(t)$ is the total output of the electric power per time unit t,

$\lambda(t)$ is price of the new elements constructed per time unit t,

$p\left(t\right)$ is the output rate for external production of the plant created by subsystem B (the known dynamics for power consumption, i.e., electric load) per time unit t.

Functions $\alpha_{1,2}$ are conventionally supposed to be non-decreasing functions in case of the production capacities' strict lifetime. This condition can be relaxed as demonstrated below and in the 1st Part of this book.

It is to be noted here that for closure of the system (12.2.1)–(12.2.6) one must know the functions $x^0(t)$, $y^0(t)$ on the prehistory $[\underline{a}(t_0), t_0]$, where $\underline{a}(t_0) = \min(\alpha_1(t_0), \alpha_2(t_0))$. Thus, apart from (12.2.1)–(12.2.6) one must know

$$x = x^0(t), \ y = y^0(t), \ t \in [\underline{a}(t_0), t_0].$$

Let us now address the special case of the two-sector macroeconomic model (12.2.1)–(12.2.6) is case of the subsystem A absence, when $y(t) \equiv 0$ as follows

$$x\left(t\right) = \int_{a(t)}^{t} \alpha(t,s)x\left(s\right)y\left(s\right)ds + g(t), \quad t \in [t_0, T], \tag{12.2.7}$$

$$p\left(t\right) = \int_{a(t)}^{t} \beta(t,s)x\left(s\right)\left[1 - y\left(s\right)\right]ds, \tag{12.2.8}$$

$$x\left(t\right), p\left(t\right), g\left(t\right) \geq 0, \tag{12.2.9}$$

$$0 \leq y\left(t\right) \leq 1, \tag{12.2.10}$$

$$0 \leq \alpha\left(t,s\right) \leq 1, \quad 0 \leq \beta\left(t,s\right) \leq 1, \tag{12.2.11}$$

where:

$x\left(t\right)$ is the rate of creating new production capacities, i.e., the amount of new capacities per time unit at the time moment t produced by subsystem A,

$x\left(t\right)y\left(t\right)$ is the amount of new capacities per time unit at time moment t directed to subsystem A,

$x\left(t\right)\left[1 - y\left(t\right)\right]$ is the amount of new capacities per time unit at time moment t directed to subsystem B,

$\alpha\left(t,s\right)$ is the efficiency parameter for the operation of subsystem A, i.e., the share of $x(t)$ created per unit of production capacity at time moment s in A,

$\beta\left(t,s\right)$ is the efficiency parameter for the operation of subsystem B, i.e., the share of $x(t)$ created per unit of production capacity at time moment s in B,

$p\left(t\right)$ is the output rate for external production of the plant created by subsystem B,

$g\left(t\right)$ is the rate of input flow into the system from outside,

$a\left(t\right)$ is function of time that describes the dynamics of production capacities going out of operation:

$$a\left(t_0\right) < t_0, \tag{12.2.12}$$

It's conventionally supposed to be non-decreasing function (i.e., $a'\left(t\right) \geq 0$) in case of the production capacities' strict lifetime (this condition can be relaxed). The desired functions $x\left(t\right)$ and $y\left(t\right)$ are known on prehistory:

$$x\left(t\right) = x^0\left(t\right), \; y\left(t\right) = y^0\left(t\right), \; t \in \left[a\left(t_0\right), t_0\right). \tag{12.2.13}$$

The system (12.2.7)–(12.2.13) consists of the VIEs. In case of one-sector macroeconomic model $(y(t) \equiv 0)$ we get the VIE (2.1.1) when all the kernels except $K_n(t,s)$ are zeros. The theory of such equations with piecewise continuous kernels is studied in Part 1 of this book.

It is to be noted that [Karaulova and Markova (2003, 2008)] used the single-product models of development of the EPS with various degrees of aggregation by types of the power plants. For detailed description readers may refer to papers by [Karaulova and Markova (2008); Ivanov *et al.* (2006)].

As it has been mentioned above the desired functions must be known on prehistory. It make sense to address the following evolutionary (Volterra) equations of the first kind with piecewise continuous kernel:

$$\int_0^{a_1(t)} \mu_1(t,s)x(s)ds + \int_{a_1(t)}^{a_2(t)} \mu_2(t,s)x(s)ds$$

$$+ \ldots + \int_{a_{n-1}(t)}^{t} \mu_n(t,s)x(s)ds = p(t), \quad t \in [0,T], \qquad (12.2.14)$$

where the different parts of the system life has various intensity factors of capacities utilization $\mu_i(t,s)$.

Here for the solution's uniqueness of the equation (12.2.14) it's necessary to know $p(t)$ in the initial time moment, namely $p(0) = 0$.

12.2.2　*Numerical Solution of the Volterra Equations with Piecewise Continuous Kernels*

For sake of clarity let us start the numerical study of the VIE (2.1.1) for $n = 2$:

$$\int_0^{a(t)} K_1(t,s)x(s)ds + \int_{a(t)}^{t} K_2(t,s)x(s)ds = f(t), \ t \in [0,T], \qquad (12.2.15)$$

where $0 < a(t) < t \ \forall t \in (0,T]$, $a(0) = 0$, functions $K_1(t,s)$, $K_2(t,s)$, $f(t)$ are continuous and smooth enough, $f(0) = 0$, $K_2(t,t) \neq 0 \ \forall t \in [0,T]$. We use the method of right rectangles. For $T \in \mathbb{N}$ let us introduce the mesh $t_i \ i = \overline{1,n}$, $nh = T$ and construct a numerical approximation to (12.2.15) by the right rectangle formula as follows

$$h \sum_{j=1}^{l-1} K_1(t_i,t_j)x^h(t_j) + (a(t_i) - t_{l-1})K_1(t_i,a(t_i))x^h(a(t_i))$$

$$+ (t_l - a(t_i))K_2(t_i,t_l)x^h(t_l)$$

$$+ \ldots + h \sum_{j=l+1}^{i} K_2(t_i,t_j)x^h(t_j) = f(t_i), \quad i = \overline{1,n}, \qquad (12.2.16)$$

Table 12.3 Errors
for the test example
12.3

h	ε
1/128	0,1540369
1/256	0,0770624
1/512	0,0385212

where $l = \left[\frac{a(t_i)}{h}\right] + 1$. The appearance of the terms that are not under the sign of the sum due to the fact that the limit of $a(t_i)$ in the general case does not fall on a mesh node.

It is already for $n = 1$ is necessary to solve the equation with two unknowns: $x^h(a(t_1))$ and $x^h(t_1)$. The same problem occurs at every step (except for special cases when $a(t_i)$ falls on a mesh). To solve this problem we have the various methods. For example, combining the techniques of left and right rectangles, the use of interpolation and extrapolation procedures, and use of the knowledge of the solution of equation (12.2.15) at the initial point, see paper of [Markova and Sidorov (2014)]. The numerical computations have demonstrated the linear convergence of designed numerical method.

Example 12.1.

$$2 \int_0^{\sin \frac{t}{2}} x(s)ds + \int_{\sin \frac{t}{2}}^t x(s)ds = \frac{1}{3}\sin^3 \frac{t}{2} + \frac{t^3}{3}, \ t \in [0, 2\pi],$$

exact solution $\bar{x}(t) = t^2$.

In Table 12.4 the errors $\varepsilon = \max_{1 \leq i \leq n} |\bar{x}(t_i) - x^h(t_i)|$ are listed for various steps, Fig. 12.2 represents the error behavior for the step $h = \frac{\pi}{64}$ for the major (solid) and for the minor (dotted) mesh nodes. Here minor mesh consists of nodes that are not the sum in (12.2.16).

Let us now describe the generic numerical method for Volterra integral equations (2.1.1) with piecewise continuous kernel (2.1.2) based on the mid-rectangular quadrature rule recently proposed by [Sidorov et al. (2014)]. The accuracy of proposed numerical method is $\mathcal{O}(1/N)$.

For numerical solution of the equation (2.1.1) on the interval $[0, T]$ we

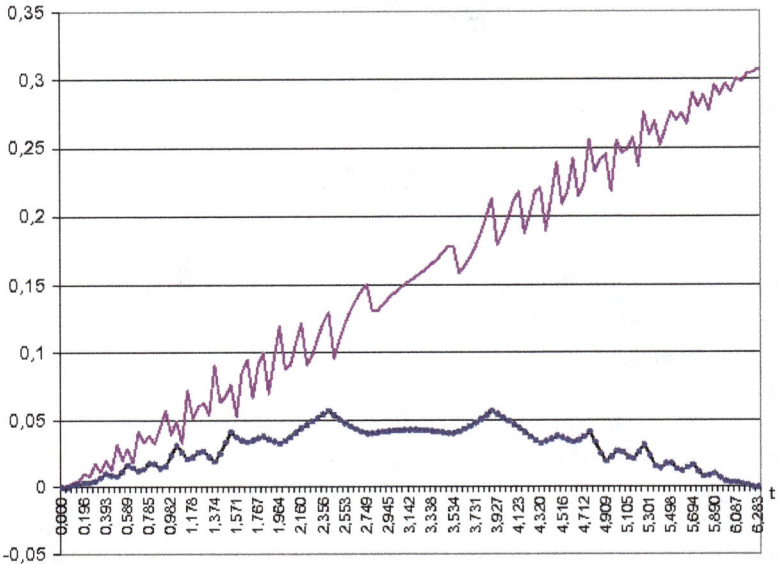

Fig. 12.2. Errors behavior.

introduce the following mesh (the mesh can be non-uniform)

$$0 = t_0 < t_1 < t_2 < \ldots < t_N = T, \quad h = \max_{i=\overline{1,N}}(t_i - t_{i-1}) = \mathcal{O}(N^{-1}).$$

$$(12.2.17)$$

Let us search for the approximate solution of the equation (2.1.1) as follows

$$x_N(t) = \sum_{i=1}^{N} x_i \delta_i(t), \ t \in (0, T], \ \delta_i(t) = \begin{cases} 1, \text{ for } t \in \Delta_i = (t_{i-1}, t_i]; \\ 0, \text{ for } t \notin \Delta_i \end{cases}$$

$$(12.2.18)$$

with coefficients x_i, $i = \overline{1, N}$ are under determination.

In order to find $x_0 = x(0)$ we differentiate both sides of the equation (2.1.1) wrt t:

$$f'(t) = \sum_{i=1}^{n} \left(\int_{\alpha_{i-1}(t)}^{\alpha_i(t)} \frac{\partial K_i(t, s)}{\partial t} x(s) ds + \alpha_i'(t) K_i(t, \alpha_i(t)) x(\alpha_i(t)) \right.$$

$$\left. - \alpha_{i-1}'(t) K_i(t, \alpha_{i-1}(t)) x(\alpha_{i-1}(t)) \right).$$

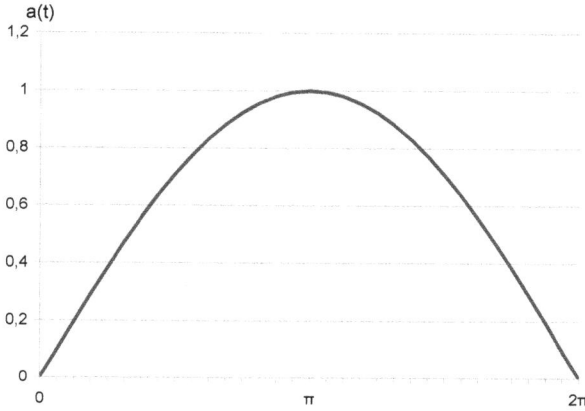

Fig. 12.3. $a(t) = \sin \frac{t}{2}$.

Therefore

$$x_0 = \frac{f'(0)}{\sum\limits_{i=1}^{n} K_i(0,0) \left[\alpha'_i(0) - \alpha'_{i-1}(0)\right]}. \tag{12.2.19}$$

Here we assume that conditions of the Theorem 2.4.7 are satisfied and $\sum\limits_{i=1}^{n} K_i(0,0) \left[\alpha'_i(0) - \alpha'_{i-1}(0)\right] \neq 0$.

Let's make the notation $f_k := f(t_k)$, $k = 1, \ldots, N$. In order to define the coefficient x_1 we rewrite the equation in $t = t_1$:

$$\sum_{i=1}^{n} \int_{\alpha_{i-1}(t_1)}^{\alpha_i(t_1)} K_i(t_1, s) x(s) ds = f_1. \tag{12.2.20}$$

It is to be noted that the lengths of all the segments of integration $\alpha_i(t_1) - \alpha_{i-1}(t_1)$ in (12.2.20) are less or equal to h. Then application of the mid-rectangular quadrature rule yields

$$x_1 = \frac{f_1}{\sum\limits_{i=1}^{n} (\alpha_i(t_1) - \alpha_{i-1}(t_1)) K_i(t_1, \frac{\alpha_i(t_1) + \alpha_{i-1}(t_1)}{2})}. \tag{12.2.21}$$

The mesh point of the mesh (12.2.17) which coincide with $\alpha_i(t_j)$ we denote as v_{ij}, i.e., $\alpha_i(t_j) \in \Delta_{v_{ij}}$. Obviously $v_{ij} < j$ for $i = \overline{0, n-1}$, $j = \overline{1, N}$. It is to be noted that $\alpha_i(t_j)$ are not always coincide with any mesh point. Here v_{ij} is used as index of the segment $\Delta_{v_{ij}}$, such as $\alpha_i(t_j) \in \Delta_{v_{ij}}$ (or its right-hand side).

Let us now assume the coefficients $x_0, x_1, \ldots, x_{k-1}$ be known. Equation (2.1.1) defined in $t = t_k$ as

$$\sum_{i=1}^{n} \int_{\alpha_{i-1}(t_k)}^{\alpha_i(t_k)} K_i(t_k, s)x(s)ds = f_k,$$

we can rewrite as follows: $I_1(t_k) + I_2(t_k) + \cdots + I_n(t_k) = f_k$, where

$$I_1(t_k) := \sum_{j=1}^{v_{1,k}-1} \int_{t_{j-1}}^{t_j} K_1(t_k, s)x(s)\,ds + \int_{t_{v_{1,k}}-1}^{\alpha_1(t_k)} K_1(t_k, s)x(s)\,ds,$$

$$\ldots$$

$$I_n(t_k) := \int_{\alpha_{n-1}(t_k)}^{t_{v_{n-1,k}}} K_n(t_k, s)x(s)\,ds + \sum_{j=v_{n-1,k}+1}^{k} \int_{t_{j-1}}^{t_j} K_n(t_k, s)x(s)\,ds.$$

(1) If $v_{p-1,k} \neq v_{p,k}$, $p = 2, \ldots, n-1$ then

$$I_p(t_k) := \int_{\alpha_{p-1}(t_k)}^{t_{v_{p-1,k}}} K_p(t_k, s)x(s)\,ds + \sum_{j=v_{p-1,k}+1}^{v_{p,k}-1} \int_{t_{j-1}}^{t_j} K_p(t_k, s)x(s)\,ds$$

$$+ \int_{t_{v_{p,k}}-1}^{\alpha_p(t_k)} K_p(t_k, s)x(s)\,ds.$$

(2) If $v_{p-1,k} = v_{p,k}$ then

$$I_p(t_k) := \int_{\alpha_{p-1}(t_k)}^{\alpha_p(t_k)} K_p(t_k, s)x(s)\,ds.$$

Remark 12.1. The number of terms in each line of the last formula depends on an array v_{ij}, defined using the input data: functions $\alpha_i(t)$, $i = \overline{1, n-1}$, and fixed (for specific N) mesh.

Each integral term we approximate using the mid-rectangular quadrature rule, e.g.,

$$\int_{t_{v_{p,k}}-1}^{\alpha_p(t_k)} K_p(t_k, s)x(s)ds$$

$$\approx (\alpha_p(t_k) - t_{v_{p,k}-1}) K_p\left(t_k, \frac{\alpha_p(t_k) + t_{v_{p,k}-1}}{2}\right) x_N\left(\frac{\alpha_p(t_k) + t_{v_{p,k}-1}}{2}\right).$$

Moreover, on those intervals where the desired function has been already determined, we select $x_N(t)$ (i.e., $t \leqslant t_{k-1}$).

On the rest of the intervals an unknown value x_k appears in the last terms. We explicitly define it and proceed in the loop for k. The number of these terms is determined from the initial data v_{ij} analysis. The accuracy of the numerical method is $\mathcal{O}\left(\frac{1}{N}\right)$.

Let us consider three examples to illustrate the efficiency of proposed numerical method. In all the examples below the uniform mesh is employed. Tables demonstrates the errors $\varepsilon_N = ||x^N(t_i) - \bar{x}(t_i)||_{\Omega^N}$ and order of convergence $p^N = \log_2 \frac{D^N}{D^{2N}}$ based on maximum pointwise two-mesh differences $D^N = ||x^N(t_i) - x^{2N}(t_i)||_{\Omega^N}$ without *a priori* knowledge of exact solution.

Example 12.2.

$$\int_0^{t/3} (1+t-s)x(s)\,ds - \int_{t/3}^t x(s)\,ds = \frac{t^4}{108} - \frac{25t^3}{81},$$

exact solution is $\bar{x}(t) = t^2$, $t \in [0, 2]$. Table 12.4 demonstrates the errors and order of convergence analysis.

Table 12.4 Errors, two-mesh differences and order of convergence for the 1st Example.

	32	64	128	256	512	1024	2048	4096
ε_N	0.13034	0.07804	0.03989	0.01975	0.01002	0.00508	0.00256	0.00128
D^N	0.07462	0.03815	0.02013	0.00975	0.00514	0.00259	0.00129	0.00065
p^N	0.96774	0.92207	1.04619	0.92217	0.98864	1.00716	0.98639	1.00198

Example 12.3.

$$\int_0^{\frac{t}{9}} (1+t-s)x(s)\,ds - \int_{\frac{t}{9}}^{\frac{2t}{9}} x(s)\,ds - 2\int_{\frac{2t}{9}}^{\frac{4t}{9}} x(s)\,ds + \int_{\frac{4t}{9}}^t x(s)\,ds = \frac{11t^4}{26244} + \frac{547t^3}{2187},$$

exact solution is $\bar{x}(t) = t^2$, $t \in [0, 2]$. Table 12.5 demonstrates the errors ε_N, order of convergence p^N and maximum pointwise two-mesh differences D^N for various N.

Table 12.5 Errors, two-mesh differences and order of convergence for the 2nd Example.

	32	64	128	256	512	1024	2048	4096
ε_N	0.13718	0.07408	0.04531	0.02211	0.01107	0.00549	0.00274	0.00141
D^N	0.09030	0.04638	0.02605	0.01413	0.00689	0.00346	0.00169	0.00092
p^N	0.96113	0.83188	0.88279	1.03428	0.99528	1.02747	0.86853	0.94392

Example 12.4.

$$2\int_0^{\sin\frac{t}{2}} x(s)\,ds - \int_{\sin\frac{t}{2}}^{2\sin\frac{t}{3}} x(s)\,ds + \int_{2\sin\frac{t}{3}}^{t} x(s)\,ds = \frac{t^3}{3} + \sin^3\frac{t}{2} - \frac{16}{3}\sin^3\frac{t}{3},$$

exact solution is $\bar{x}(t) = t^2$, $t \in \left[0, \frac{3\pi}{2}\right]$. Table 12.6 demonstrates the errors ε_N, order of convergence p^N and maximum pointwise two-mesh differences D^N for various N.

Table 12.6   Errors, two-mesh differences and order of convergence for the 3rd Example.

	32	64	128	256	512	1024	2048	4096
ε_N	0.74659	0.36399	0.21969	0.11931	0.10206	0.06504	0.05814	0.05149
D^N	0.43430	0.29149	0.15958	0.15915	0.15874	0.12170	0.09866	0.09006
p^N	0.57524	0.86911	0.00393	0.00372	0.38329	0.30276	0.13158	0.03375

Chapter 13

Suppression of Moiré Patterns for Video Archive Restoration

13.1 Introduction. Chapter Summary

In this chapter we address the employment of the Fourier integral transform in digital signal processing application.

The work presented in this chapter was supported by the EU's IST research and technological development programme, the results are obtained in collaboration with A. Kokaram (Trinity College Dublin). It is carried out within the BRAVA project ("Broadcast Restoration Through Video Analysis"). The reader may see the papers of [Kokaram *et al.* (2002); Sidorov and Kokaram (2002)] for more details regarding the BRAVA project.

The moiré effect is a phenomenon which occurs when two or more images with periodic or quasi-periodic structures (such as dot screens, line gratings, etc.) are non-linearly combined to create a new superposition image. Moiré patterns do not exist in any of the original images, but appear in the superposition image as result of a multiplicative superposition rule. However, the multiplicative model is not the only possible superposition rule. In other situations a different interference rule can be appropriate. In the restoration film archives we deal with a multiplicative superposition rule caused by Telecine processing. This will be discussed more it details in next section. For more details on motion pictures restorations readers may refer to the monograph of [Kokaram (1998)].

While the moiré phenomenon has useful applications in several fields, such as in strain analysis or in the detection and measurement of slight deflections or deformations (see publications of [Oster (1964); Theocaris (1963); Takasaki (1970); Nishijima and Oster (1964)]), in many areas, moiré patterns have an unwanted, adverse effect. In the digital image processing literature moiré effects are often considered, in view of sampling theory, as aliasing phenomena. Sampling moiré mainly occur in the analog to digital conversion process between repetitive patterns in the original image and the device sampling lattice. It is can be considered as a special case of superposition moiré as outlined in the book of [Amidror (2000)]. It is to be noted that moiré removal is inverse and ill-posed problem. For more details concerning the theory of well-posed, ill-posed, and intermediate problems and relation to image reconstruction readers may refer to the book of [Petrov and Sizikov (2005)].

In this chapter we consider the problem of sampling moiré filtering of motion pictures.

13.2   Moiré Degradation on Film

In the archive restoration process, the moiré phenomenon is an optical effect, which can appear during the Telecine transfer due to the difference between the orientation of TV lines captured on film and the scanning angle of the Telecine. Figure 13.1 shows a block diagram outlining the elements in the production chain that may result in the formation of moiré. Before the use of video recording machines using magnetic tape, a device called a Kinescope was used to record the pictures on the Television monitor directly onto film. The Television scan lines are observed to be transferred directly to the film and this is illustrated in the Fig. 13.1. The spacing and geometry of the scan lines depends on the optical projection in the Kinescope as well as the geometry of the old Television monitors. Those monitors were far from flat. Thus the Television scan lines transferred to the film would have curvilinear distortion as well as tilt due to the raster of the Television set.

In order to convert the film recording of the Television event into a modern broadcast format, the film must now be re-scanned in the Telecine device. However, the Telecine uses a flying spot to scan the film (here readers may refer to the book of [Kallenberg and Cvjetnicanin (1994)]) and the orientation and spacing of this scan pattern will not necessarily align with the original Television line scan. This will cause aliasing in the resulting signal, and that manifests as a periodic pattern superimposed on the resulting image. To circumvent this problem, it is necessary to "blur" or change the focus of the flying spot so that the aliasing is reduced.

Unfortunately, this is not often simple to do and the resulting image is corrupted with moiré. A frame with moiré degradation is shown in Fig. 13.2. The moiré manifests here as periodic dark bands that become

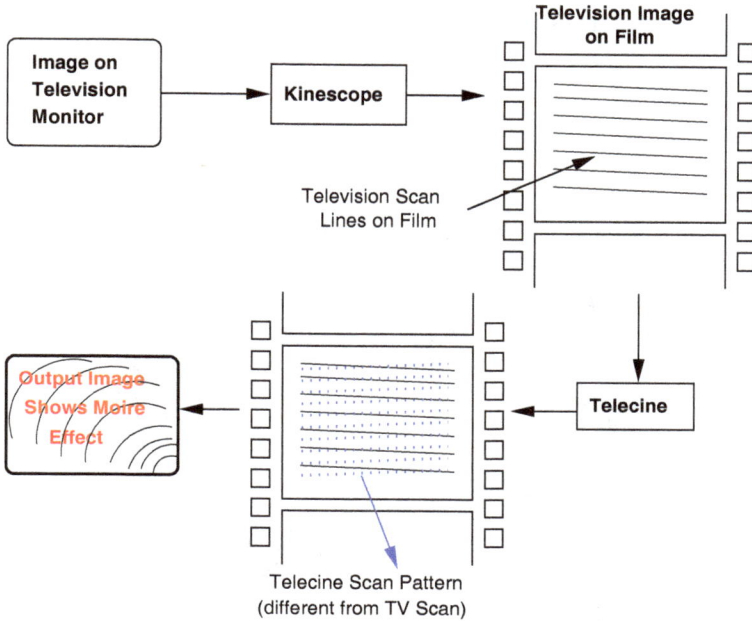

Fig. 13.1.   A block diagram of the system causing moiré degradation

curved toward the extremities of the image. Intensity and curvilinear nature of moiré patterns is highly non-stationary and non-regular. We suspect that the curved nature of the bands is as a direct result of the curved geometry of the original Television monitors.

13.3   Previous Work. The State of the Art

Several mathematical approaches in terms of the inverse problem theory in the image domain have been used to explore moiré phenomenon. The pure algebraic approach, which is based on the geometric properties of superposed periodic layers, was developed by [Oster *et al.* (1964)] in 1964. In the field of color reproduction moiré patterns may seriously deteriorate the image quality. An algorithm for moiré minimization for color printing was presented in the book of [Amidror (2000)] in the monograph "The theory of the moiré phenomenon". Most of the moiré minimization methods discussed there presume preprocessing operations (i.e., pre-sampling, re-scunning) which are not available in the problem of archive moiré removal.

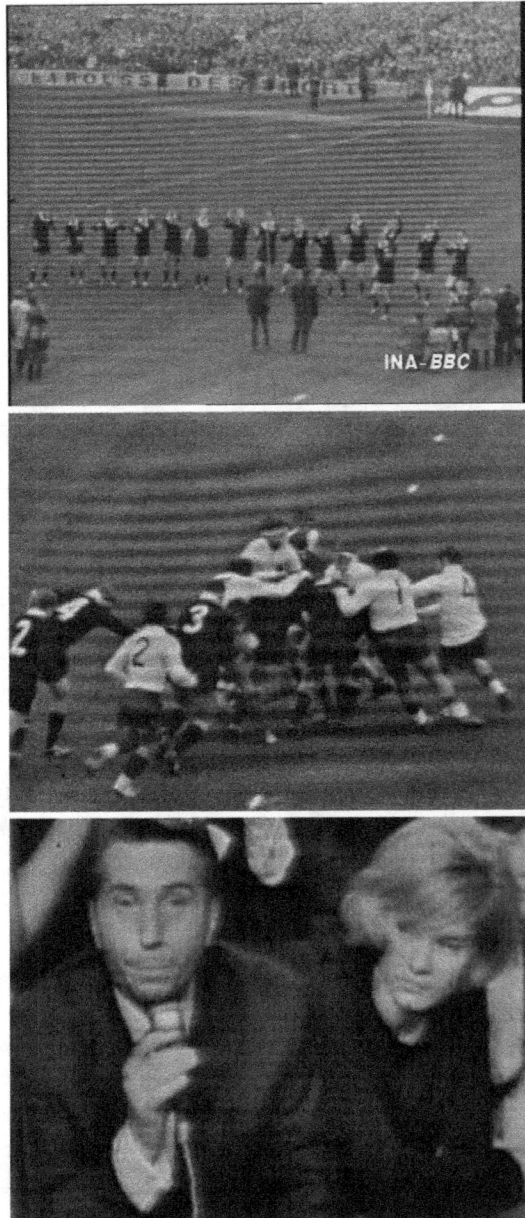

Fig. 13.2.   Frames degraded with moiré interferences.

A video moiré cancellation non-linear filter for high-resolution CRTs was described by [Hentschel (2001)]. It was shown that such an approach fails in moving images and manual parameter adjustment remains difficult. A quasi-periodic noise suppression algorithm was applied for satellite photographs' enhancement by [Yaroslavsky and Eden (1996)] in 1979. This algorithm is based on notch filtering by Hadamar spectrum analysis. A possible direct approach [Cho *et al.* (1989)] for elimination of 2D periodical interference pattern superimposed on an image is to manually remove a pair of symmetric peak impulses in Fourier Spectrum which correspond to interference. This kind of approach is appropriate for off-line image processing only. In general case it is impossible to manually remove quasi-periodic interferences.

Review of 2D digital notch filters for additive periodic noise suppression was given by [Hinamoto *et al.* (2000)] *et al.* and by [Pei and Tseng (1994)].

[Aizenberg and Butakoff (2002)], [Aizenberg *et al.* (2002)] designed a median-like notch filter for suppression of periodic and quasi-periodic noise in photographs. The patterns discussed there could be characterized as a stationary case of non-stationary quasi-periodic archive moiré phenomena.

The Fourier Spectrum domain peak elimination by thresholding of the amplitude spectrum is also described by [Arthur and Weeks (1996)].

The moiré phenomenon investigated here has a different basis altogether and previous work, although interesting, is not necessarily relevant to archive moiré phenomena.

The purpose of this chapter is to demonstrate the efficiency of integral transform for non-stationary signal processing via presentation of the automated 2D notch filter design techniques for suppression of moiré interference in video archives.

13.4   Model of the Archive Moiré Phenomena

Consider first of all a 1D description of the phenomenon. Say that the signal stored on our 1D film consists of single isolated intensities. These intensities would coincide with the Television scan lines stored on the film. Let this signal be $y(x)$ and the spacing between the intensities be T_1. x is a continuous position coordinate in the film.

Then we could define our analogue signal by

$$y(x) = \sum_{n=0}^{N-1} I_n \delta(x - nT_1). \tag{13.4.1}$$

This signal is shown at the top of Fig. 13.3. T_1 therefore models the spacing between the scan lines on the film, and I_n is the intensity of the Television image in each line, but in this 1D case the line is just a point.

We can decompose the action of the flying spot in the Telecine as first a blurring (low pass filter) operation on $y(x)$ followed by a resampling of the resulting signal. Assuming a low pass filter $g(x)$, for instance a Gaussian filter, the low passed signal $y_l(x)$ is given by the convolution integral

$$y_l(x) = y(x) * g(x). \tag{13.4.2}$$

The low passed signal in this case is illustrated by the second row in the Fig. 13.3.

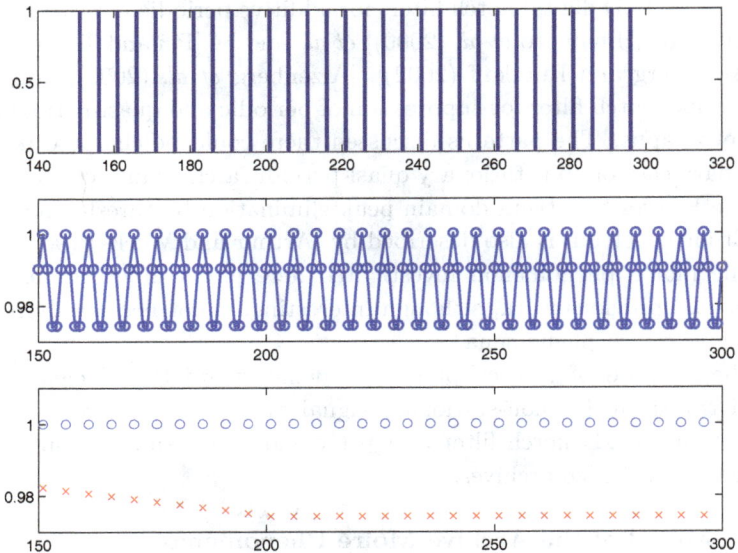

Fig. 13.3.   **Top:** Original 1D signal stored on film. **Middle:** Blurring caused by Flying Spot. **Bottom:** Resampled signal not the same as original.

The last step is a sampling step using sampling period $T_2 = \alpha T_1$ where α is close to 1. This models the fact that the Telecine scanning period is close to that required, but not quite right. T_2 is the spacing between the Telecine scan lines.

We could introduce as well a phase shift or offset to the resampling operation, but we omit this for the sake of clarity here. Thus our final sampled signal y_k is given by

$$y_k = y_l(kT_2) = \int_\tau y(\tau)g(kT_2 - \tau)d\tau . \qquad (13.4.3)$$

Using $T_2 = 0.99T_1$, and a Gaussian blur with a 20 tap window having a variance of 6.25, the final sampled signal in our example is shown as the bottom row of the Fig. 13.3 and in the Fig. 13.4. As can be seen the final signal has a superimposed periodic pattern. This is the moiré degradation being addressed. The Fig. 13.4 shows more of the periodic sampled signal and the original signal. The resampled signal is clearly affected by a periodic pattern that is the moiré.

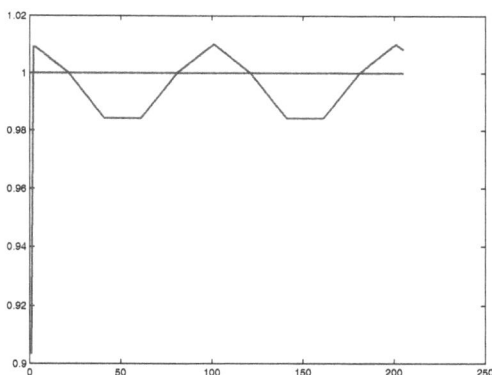

Fig. 13.4. The straight line is the true signal, and the periodic signal is the moiré effect.

Consider a 2D description of the phenomenon. The idea is the same. In Television monitors, the diameter of the scanning spot is less than the gap between adjacent lines. This means that there is a strip of darkness (black) between each line. First, the distance between Television scan lines modelled by inserting strips of black between each lines of the original image. Then the spread of the scanning dot in the Telecine device is created by applying a Gaussian filter as shown above in 1D case. Finally, we produce resampling of the image by interpolation between adjacent Television scan lines. It is the ratio between the number of inserted lines and the spread of the Gaussian blur that is equivalent to the line spacing in the Telecine.

Figures 13.5 and 13.6 show Lenna artificially corrupted with moiré using $\sigma = 0.6$ (and 9×9 taps) for the Gaussian blur and $\alpha = 0.9$. In this case, only one line was inserted between each line of the original data. The right

Fig. 13.5. Vertical misalignment in the Telecine.

Fig. 13.6. Vertical and Rotational misalignment in the Telecine.

hand side figure shows the effect of a rotational misalignment in the Telecine scanning pattern using an angle of $\theta = 0.5°$. In real video sequences there are two slightly different moiré patterns in odd and even Television scan lines which is the effect of Telecine device scanning dot technology. The difference between these patterns are difficult to see in stills (see Fig. 13.7).

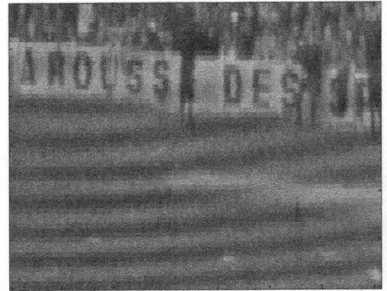

Fig. 13.7. Patch of degraded frame. **Left:** Odd lines. **Right:** Even lines.

This model of periodic interference is not the only possible. But even if visually such quasi-period non-stationary distortion looks appropriate, we cannot rely on this model because different interference rules in the image domain will have different spectra composition rules in spectral domain.[1]

---

[1]Let us outline here that image reconstruction algorithm in case of noisy point-spread function was proposed by [Voskoboinikov (2007)] in terms of regularization of the integral equation of the first kind. Here readers may also refer to the monograph of Petrov, Y. P. and Sizikov, V. S. (2005).

13.5 Moiré Removal

From the model and examples above we can consider that the moiré is a sinusoidal degradation caused by improper sampling in the Telecine device. In that particular example, the degradation can be modelled as either additive or multiplicative. Regardless of that observation it is the non-stationary geometry of the pattern that remains difficult to model.

The natural approach for the sinusoidal noise filtering is frequency domain processing. We designed an adaptive spectral filter which parameters are adjusted to s a unit gain on all frequencies except for moiré frequency where the gain must be zero. This will be discussed next.

13.5.1 *Spectral Moiré Suppression*

The essence of the idea is to detect peaks in the spectrum of the degraded image and delete these peaks assuming that they represent the moiré. The cosine and Hadamar spectra are also suitable for the detection of sinusoidal disturbance. The Walsh transform is acceptable for suppression of rectangular wave noise. This scheme exploits the two-dimensional Fourier integral transform . It is based on analysis of 2D Fast Fourier transform (FFT) spectra.

The steps of the spectrum modification are as follows

(1) *Protect a band of low frequencies in the image from manipulation* (see Fig. 13.9). Assuming that the moiré is a predominantly horizontal periodic pattern, the peaks in the frequency space will appear near the vertical frequency axis. Define a range $\pm\Omega$ of vertical frequency within which the spectrum will be untouched because the most of the energy of the image is accumulated in the low-frequency coefficients that have much higher amplitudes in comparison with medium and high frequency spectral coefficients where moiré peaks are normally appear.

(2) *Create a detection field* $B(\omega_h, \omega_v)$ which is set to 1 at frequency components that should be removed and 0 otherwise, as follows

$$B(\omega_h, \omega_v) = \begin{cases} 0 & \text{If } (|F(\omega_h, \omega_v)| > \Delta) \text{ and } (|\omega_v| > \Omega) \\ 1 & \text{Otherwise} \end{cases} \tag{13.5.1}$$

where $|F(\omega_h, \omega_v)|$ is the magnitude of the spectral component at frequency (ω_h, ω_v), the threshold Δ is selected by analysis of the cumulative probability distribution of spectral components. We detect top 10% of $|F(\omega_h, \omega_v)|$, for $|\omega_v| > \Omega$. The typical CPDF of the spectrum coefficients is shown in the Fig. 13.8 for the frame shown in the Fig. 13.17.

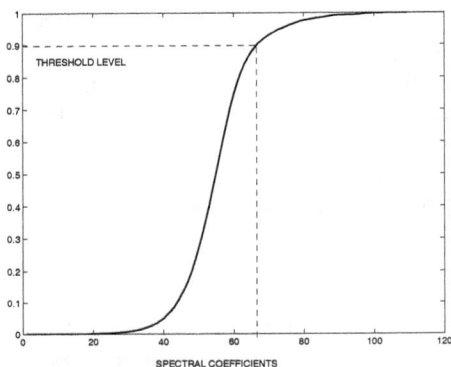

Fig. 13.8.   CPDF of the spectral coefficients. Dashed line shows the threshold's selection procedure.

The choice of Ω is supposed to be done according the degradation level of the sequence. Our experiments show that $\Omega \in (5, 15)$ covers a wide range of different moiré intensities for the standard TV resolution frames (720×576).

(3) *Replace the magnitude of each spectral component identified for removal by* $B(\cdot, \cdot) = 0$, *with a noise floor.* This noise floor value is a local median of the magnitude of the high frequency spectral coefficients in ε-neighborhood (window) around the vertical DC axis (DC area is not included) as it shown in the Fig. 13.9. Experiments show conventional results for $\varepsilon = 10$. It is to be noted that high frequency spectral components do not contain as much information as low frequency components. That is explains why this kind of ad-hoc operations give stable conventional results.

Figure 13.17 shows the results of this process operating on a real degraded image frame for $\Delta = 94.4$ and $\Omega = 25$. Much of the moiré is removed, but the image is corrupted by a strong ringing effect.

The ringing effect appears in the image domain because of Gibbs phenomenon. To illustrate the cause of the effect in the frequency domain, the

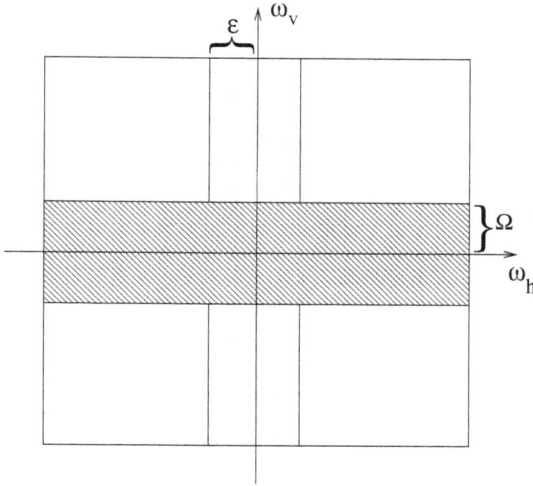

Fig. 13.9.   Diagram of 2D Fourier Spectrum Domain: crosshatched region shows filter pass band, gray color region shows moiré area.

Fig. 13.14 shows a slice in the 2D spectrum of the original, corrupted and restored Lenna at zero horizontal frequency. The ringing effect (easier to see in a sequence rather than stills) is caused by discontinuities inserted in the Fourier domain at the location of the moiré frequencies. The plateaus that were inserted are shown in the Fig. 13.14.

The presence of the slight ringing effect could imply that the distortion is better removed in the image domain itself. This will be discussed in next section.

13.5.2   *Overlapped Processing*

In order to minimize the ringing effect we may use two approaches. One approach is to implement a soft thresholding operation in the spectral domain. This however, does not solve the problem entirely and it transpires that in fact, the greater problem is that the moiré pattern appears to be frequency modulated in both image directions. The other approach is to use overlapped blocks for processing. Within each block, the moiré pattern is then more regular and yields to this type of spectral processing.

The choice of analysis and synthesis windows for overlapped processing is important. The analysis window must allow good spectral resolution, while the synthesis window must allow the output data to overlap such

Fig. 13.10. Fourier spectrum of degraded image.

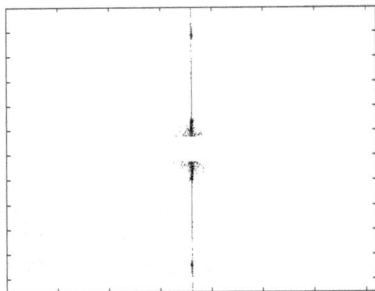

Fig. 13.11. Frequencies marked for removal.

Fig. 13.12. Image degraded with moiré.

Fig. 13.13. Restored image.

that the overlapped portions of data show a net gain of unity through the system. The reader may see the monograph of [Kokaram (1998)] for a treatment of overlapped processing for the noise reduction in images. In this work, we have found that vertically overlapping image horizontal stripes by 2:1 provides good results since the moiré pattern is primarily horizontal.

The analysis and synthesis windows are the same and the net effect is that of using a raised cosine window as follows

$$h_n = \cos^2\left(\frac{\pi n}{2N}\right),$$

$$\text{for } n = -\left(N - \frac{1}{2}\right), -\left(N - \frac{1}{2}\right) + 1, \ldots, \left(N - \frac{1}{2}\right).$$

(13.5.2)

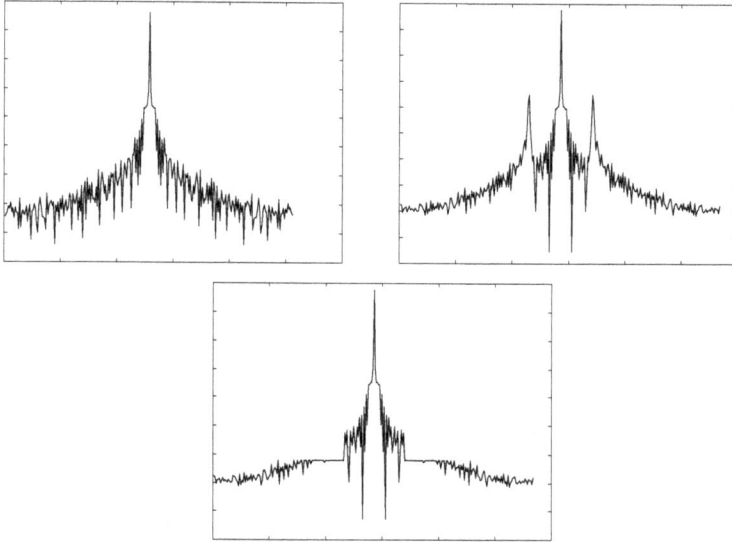

Fig. 13.14. A slice in the magnitude of the 2D spectrum of Lenna at zero horizontal frequency. **Left:** Original image. **Right:** Corrupted image. **Bottom:** Restored Image.

They are both even length windows of N taps. The windows are applied vertically only since the overlapped blocks are in fact stripes that extend over the whole image width. The algorithm is then as follows

(1) Split the frame into even length horizontal stripes that are over-lapped by half their height.
(2) Window each stripe by the analysis window (applied to each column) in the vertical direction.
(3) Apply the spectrum domain based non-linear filter (above) to each of the stripes.
(4) Window the output stripe using the synthesis window (on each column).
(5) Sum the overlapped stripes to generate the final processed image.

Figure 13.15 shows a block diagram of the system. Our restoration scheme provides ringing effect free results. Figure 13.16 shows the results of the overlapped processing algorithm operating on a real degraded image frame for real video sequence. In this example we use 128 pixel high stripes, and $\Omega = 5$. From Figs. 13.16 and 13.17 we can clearly see that overlapped processing algorithm provides visibly better results.

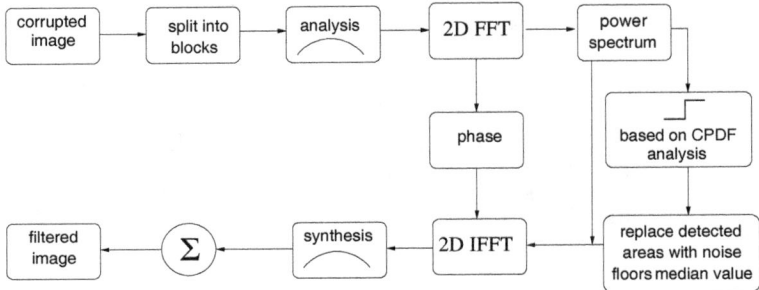

Fig. 13.15.   Block diagram of a spectrum domain based filter for suppression of moiré.

We have tested the algorithm for different levels of moiré brightness using the model (13.4.1)–(13.4.3). We can see that our algorithm provides better results for moiré patterns with higher intensity. It is to be noted that the intensity of moiré patterns remains temporally stable in video archives' materials.

Figure 13.18 shows a slice in the 2D spectrum of the corrupted and restored frame of Rugby sequence at zero horizontal frequency. We can see that there are no outlying peaks in the spectra after filtering. Figure 13.19 demonstrates a part of 2D spectrum before and after filtering.

13.6   Final Comments and Conclusions

In this section we have discussed removal of undesirable moiré interferences, which may occur in the analog to digital conversion of film recordings of Television. A mathematical integral model of moiré distortion in video was presented and shown to produce convincing moiré patterns. It is to be noted that the tests on 1D signals show indications that the moiré effect is not additive but multiplicative and cannot be modelled as Gaussian process.

It should be noted, that this model is not the only possible, and in other situation (not only in video archive restoration) different superposition rules can be appropriate. Different superposition rules in the image domain will have different spectra composition rules in spectral domain, which are determined with Fourier transform's properties.

This explains why moiré removal is fundamentally difficult. In this case the problem is anti-aliasing after the aliasing has occurred. It could be the case that some of the work in super-resolution may be relevant here, but

Fig. 13.16.   Patch of the frame filtered with $\Omega = 12$. **Top:** Degraded frame. **Middle:** Result of restoration. **Bottom:** Suppressed moiré patterns.

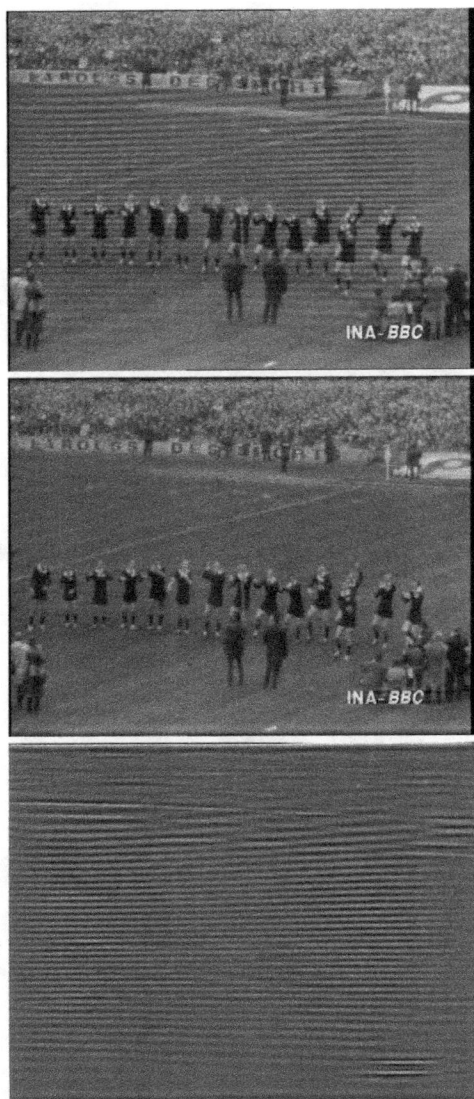

Fig. 13.17. Frame filtered with $\Omega = 5$. **Top:** Degraded frame. **Middle:** Result of restoration. **Bottom:** Suppressed moiré patterns.

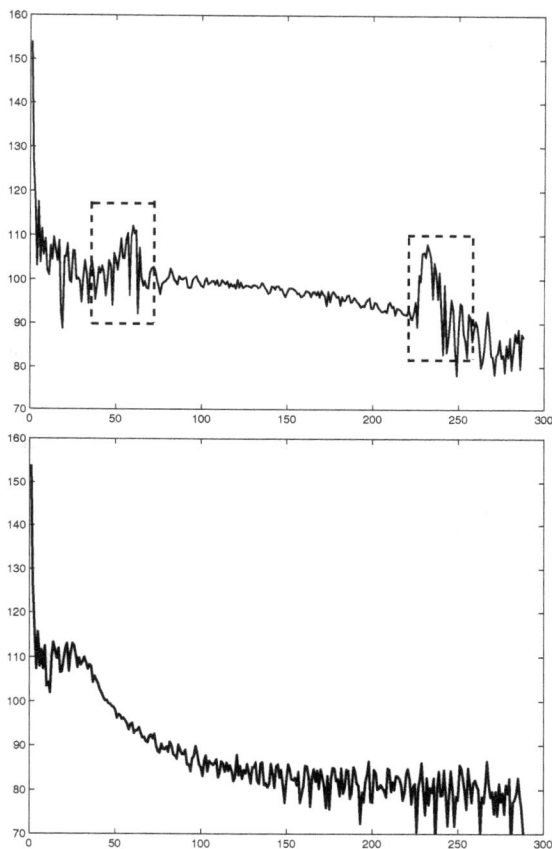

Fig. 13.18. A slice in the magnitude of the 2D spectrum of Rugby seq. frame at zero horizontal frequency. **Top:** Corrupted frame. Moiré peaks are framed with dashed line boxes. **Bottom:** Restored frame.

the displacement between the frames in a sequence may not be enough to yield more information for anti-aliasing.

Key problems are the frequency modulated aspect of the pattern and the curved geometry. The curves in the moiré are presumably due to the non-flat TV screens used in the Kinescopes of the 1950's/60's. Non-regularity of moiré interferences cause difficulties with parametric modelling of this phenomena. Another candidate for moiré removal is wavelet analysis. Our experiments show that using undecimated biorthogonal spline wavelet

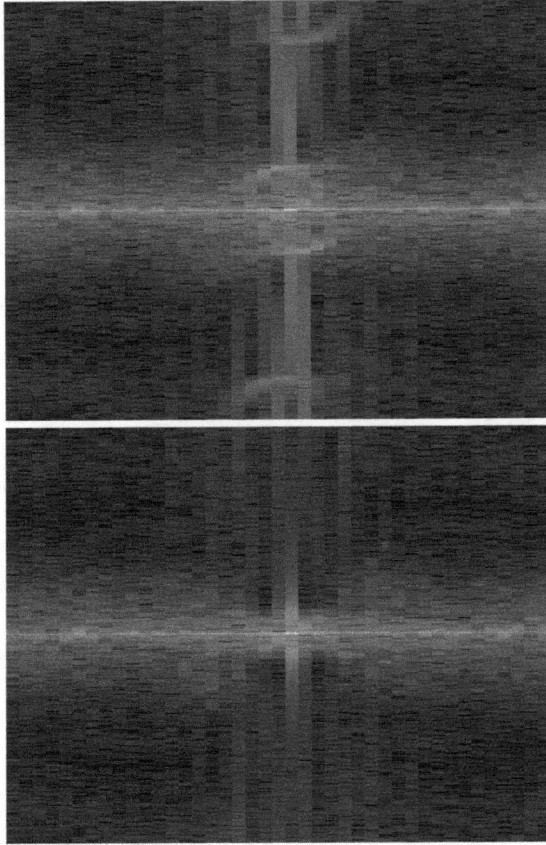

Fig. 13.19.   A part of 2D spectrum of Rugby sequence frame. **Top:** Corrupted frame. **Bottom:** Restored frame.

decomposition for moiré pattern extraction does not preserve "edges". Construction of the appropriate wavelet family remains difficult.

Some positive results could be achieved by using Wiener filter. Despite the fact that this filter gives the best linear mean square estimate of the object from the observations, it is complicate to apply it for moiré suppression. It is necessary to build a precise and universal moiré noise model, which is not easy task due to the spatial and temporal non-stationarity of this distortion.

In this chapter we discussed an effective computationally simple and fast notch filter for moiré suppression. It employed manipulation in the Fourier spectral domain and attacks directly the moiré interferences. Our algorithm exploits overlapping blocks and locates artefact peaks using automatic thresholding based on measured spectral spatial statistics. It produces visibly pleasing results. Simulation results on still pictures as well as on the real sequences demonstrate the validity of the proposed filtering method.

Chapter 14

Integral Models Applications in Electric Power Engineering

14.1 Electric Power Systems Parameters Forecasting

14.1.1 *Introduction*

The objective of this section is to demonstrate the efficiency of integral models in electric power engineering. In particular, we concentrate on non-linear time series analysis.

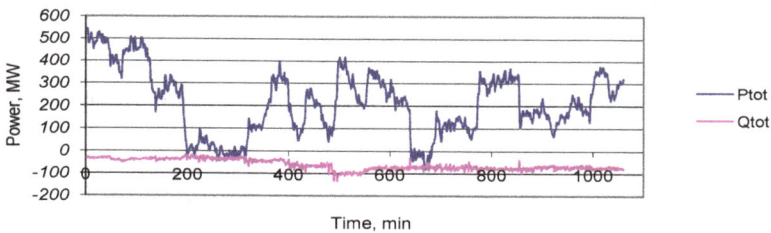

Fig. 14.1.   Change in exchange power of one TL 500kV.

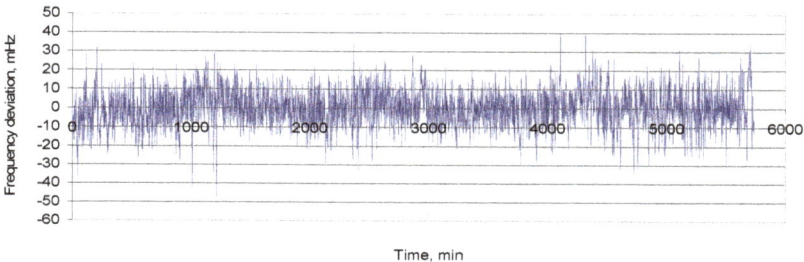

Fig. 14.2.   Typical time series: frequency deviation with "1 minute" averaging period.

The results presented in the section were obtained within the ICOEUR project concerning intelligent coordination of operation and emergency control of EU and Russian power grids. It was carried out within the Seventh Framework EU Programme. Part of the results presented in this chapter were obtained in collaboration with V. Kurbatsky, P. Leahy, V. Spyryaev, N. Tomin and A. Zhukov. For more details readers may refer to the publications of [Voropai *et al.* (2013); Kurbatsky *et al.* (2011e,a,d,c,b)].

14.1.2 *Problem Statement. Section Summary*

Forecasting of power flows, energy price, wind power etc. have become a major issue in power systems. Following the needs of today's competitive environment, various techniques are used to forecast wind power, energy price and power demand. Improvement of the efficiency of systems of emergency and secure control of electric power systems (EPS) is largely dependent on the wide-scale monitoring and forecasting the parameters of the expected operation regime of the EPS, especially it depends on the frequency of the electrical network, the voltage magnitude and the active power flows, here readers may refer to the papers of [Voropai *et al.* (2010, 2013)]. Load power in EPS varies constantly. The deviation of system load variation from the planned one leads to frequency deviations from the rated. Fig. 14.1 illustrates power change of one of two transmission lines 500 kV regulated by the power flow control system. Power deviations from the average value for one transmission line are ±200 MW at the limiting transmitted power 750 MW. It is noted that power deviations are linked not only with changes of load power, but also with restrictions on the regulated power of units electric stations.

The frequency measurements were carried out in Irkutsk EPS during four days in order to estimate the dynamics of frequency variation. The frequency deviations are shown in Fig. 14.2. Most of the deviations values lie within the limits of ±20 mHz.

Implementation of market principles in planning and control of operating conditions, expansion of the area of coordinating operation control of EPSs in terms of time (from design of control systems to their realization by dispatching and automatic devices) and situation (coordination of dispatching, continuous automatic and discrete emergency control) all cause fast dynamics of change in EPS operating conditions. The key condition for reliable work of EPS is the presence of efficient system forecasting of operation parameters (load flows, power flow, voltage magnitude, etc., see,

Fig. 14.1) and process characteristics (power and electricity losses, prices, etc). Such a forecast can prevent an abrupt change in control inputs at random nature of the frequency change caused by various malfunctions or failures of technical means of control systems, inaccuracy of input data, etc. Forecast of active power flows provides an estimation of supply ties, expected operating conditions, capacities to preserve stability of weak connection between internal and external sections of EPS by appropriate control actions. The overview of the state of the art methods for energy demand forecasting is given in the book of [Hong (2013)].

Qualitative forecasting of frequency is important for optimization of system-wide EPS management. Such forecast can prevent a sharp abrupt change of control inputs at random nature of the frequency change caused by various malfunctions or failures of technical means of control systems, inaccuracy of input data, etc. Forecast of active power flows provides an estimation of supply ties expected operation condition capacities to preserve the stability of the weak coupling of internal and external sections of EPS by issuing an appropriate control actions on regulating stations and units.

For extensive overview of recent advances concerning nonlinear time series analysis readers may refer to the book of [Small (2005)]. It is to be outlined that linear time series analysis has a long history (see book of [Vaidyanathan (2007)]) while nonlinear methods have only just begun to reach maturity.

Over the past decade, different machine learning algorithms have been investigated for analyzing power systems parameters' patterns in order to obtain forecasts for power systems with high accuracy. In forecasting the time series specified by the sharply variable non-stationary behavior, the models based on the artificial neural networks (ANN) and support vector machine (SVM) proposed by [Vapnik (1998)] became one of the most popular. Application of the SVM to power flow prediction is addressed in the paper of [Kurbatsky *et al.* (2012)]. SVM based regression (SVR) employs so called *kernel trick*, when time series modelling is addressed by extending the framework of observable linear systems to the feature space defined by a kernel which satisfies the following Mercer Theorem 14.1 from the integral equations theory (also known as the Hilbert-Schmidlt theorem). Here readers may refer to the original article of [Mercer (1909)] and

Theorem 14.1. *Suppose K is a continuous positive semi-definite kernel on a compact set X, and the integral operator $T_k : \mathcal{L}_2(X) \to L_2(X)$ defined*

by

$$T_k f := \int_X K(\cdot, x) f(x) \, dx$$

is positive semi-definite that is $\forall f \in \mathcal{L}_2(X)$,

$$\int_X K(u, v) f(u) f(v) \, dudv \geq 0.$$

Then there is an orthonormal basis $\{\psi_i\}$ *of* $\mathcal{L}_2(X)$ *consisting of eigenfunctions of* T_k *such that the corresponding sequence of eigenvalues* $\{\lambda_i\}$ *are non-negative. The eigenfunctions corresponding to non-zero eigenvalues are continuous on* X *and* $K(u, v)$ *has the representation*

$$K(u, v) = \sum_{i=1}^{\infty} \lambda_i \psi_i(u) \psi_i(v),$$

where the convergence is absolute and uniform, that is,

$$\lim_{n \to \infty} \sup_{u,v} \left| K(u, v) - \sum_{i=1}^{n} \lambda_i \psi_i(u) \psi_i(v) \right| = 0.$$

It is to be noted that the approximation performance of support vector regression is mainly based on training data and parameters selection including kernel function selection. For some results regarding the comparative analysis of the SVMs and the cluster analysis based classification methods readers may refer to the article of [Sidorov *et al.* (2008)], where the problem of automatic defect recognition and classification for vision systems development is addressed.

Let us briefly outline the main steps of the support vector regression. Let (x_i, y_i), $i = \overline{1, N}$, be the input data, namely $x_i \in \widetilde{X} \subset X, y_i \in \widetilde{Y} \subset Y$, X is the set of input data, Y is the set of values of the desired objective function. SVM resolves the regression problem by estimation of the unknown function

$$y_i = f(x_i) + \varepsilon_i, \tag{14.1.1}$$

where ε_i is error of i-th observation, e.g., in linear case $f(\mathbf{x}) = \langle \mathbf{w}, \mathbf{x} \rangle + b$, where $\mathbf{w} \in \mathbb{R}^n$ are weights and $b \in \mathbb{R}$ is the offset parameter. In nonlinear case SVM employs $\phi : \mathbb{R}^n \to H$ and instead of (14.1.1) the following function f is under the estimation $f(\mathbf{x}) = \langle \mathbf{w}, \phi(\mathbf{x}) \rangle + b$. Determination of parameters is reduced to a convex optimization problem, i.e., the following regularized risk functional minimization

$$\min_{\mathbf{w}, b, \xi, \xi^*} \left[\frac{1}{2} ||\mathbf{w}||^2 + C \sum_{i=1}^{N} (\xi_i + \xi_i^*) \right], \tag{14.1.2}$$

$$\begin{cases} y_i - \langle \mathbf{w}, \phi(\mathbf{x_i}) \rangle - b \le \varepsilon + \xi_i, \\ \langle \mathbf{w}, \phi(\mathbf{x_i}) \rangle + b - y_i \le \varepsilon + \xi_i^*, \\ \xi_i, \xi_i^* \ge 0, \ i = \overline{1, N}. \end{cases} \quad (14.1.3)$$

Here $C > 0$ is the capacity constant to trade-off between the flatness of f and the amount up to which deviations larger than ε are tolerated, ξ_i are inseparable observations parameters, ε is losses parameter. The second term in the functional (14.1.2) penalizes any deviation $f(x_i)$ from y_i based on the constraints (14.1.3). This problem is resolved in terms of the dual problem to the problem (14.1.2) and (14.1.3) as follows $f(x) = \sum_{i=1}^{N} (\alpha_i - \alpha_i^*) K(x_i, x) + b$, where $0 \le \alpha_i, \alpha_i^* \le C$, and kernel $K(x_i, x)$ satisfies the conditions of Mercer theorem 14.1 and can be represented as follows $K(x_i, x) = \langle \phi(x_i), \phi(x) \rangle$, which is normally selected as radial basis function (RBF).

Various methods of the inter-connection of neurons and organizations of their interaction can be employed to build the various ANNs. Among the collection of existing ANN structures, [Gamm *et al.* (2011)] pointed out the multilayer perceptron, the ANN on the basis of radial-basis function, and the generalized regressive network (GRNN), Volterra Neural Networks which are the most common ANNs for the short-term EPS parameters forecasting. The recurrent quadratic Volterra system to forecast the wind power output is employed in the paper of [Lee (2011)]. More details regarding the connections between the Volterra theory and polynomial regression are given in the next section. Moreover, the Volterra kernels extraction procedure has been used by [Rubiolo *et al.* (2012)] to build a Volterra-neural network (Volterra-NN) model, which is a "compressed" version of a trained multilayer perceptron (MLP) model providing the same recognition rate than the original MLP model but with fewer parameters.

One of the major problems appears in ANN trainings is the formation of an optimal input sample and parameters selection which is the challenge for times series representing non-stationary behavior of nonlinear EPS. An effective way of ANN and SVM trainings is to use nonlinear optimization algorithms, namely simulated annealing (SA) method and neuro-genetic selection of input data (NGIS), which provides a procedure for selecting the best predictive model for each sample, here readers may refer to the textbook of [Haikin (2009)].

The objective of this section is to demonstrate the efficiency of the Hilbert-Huang integral transform in power engineering for EPS parameters forecasting.

14.1.3   *Hilbert-Huang Integral Transform*

Recently, nonlinear and non-stationary analysis techniques based on the Hilbert-Huang Transform (HHT) (here readers may refer to classic article of [Huang *et al.* (1998)]) have been employed for short-term forecasting in papers of [Kurbatsky *et al.* (2011e,a,d,c)]. The Hilbert-Huang transform consists of two parts:
- Empirical Mode Decomposition (EMD);
- Hilbert Integral Transform (HT).

Let us describe the Hilbert-Huand integral transform more in details. EMD is a self-adaptive signal-processing method that is suitable for the analysis of non-linear and non-stationary processes. The main idea of the EMD lies in the decomposition of the initial signal $x(t)$ into modal basis functions with the subsequent application of the Hilbert transform to such functions. These functions obtained as result of EMD are called the intrinsic mode functions (IMFs). Thus, EMD method decomposes signal into the sum of several IMFs meeting integrity and orthogonality. As result, the signal $x(t)$ can be presented as follows:

$$x(t) = \sum_{j=1}^{n} c_j(t) + r_n(t) = \sum_{i=1}^{q} c_i(t) + \sum_{j=q+1}^{p} c_j(t)$$

$$+ \sum_{k=p+1}^{n} c_k(t) + r_n(t), \qquad (14.1.4)$$

where $q < p < n$, $c_i(t)$ are the high-frequency components, $c_j(t)$ are the components that determine the physical properties of the series, and $c_k(t)$, $r_n(t)$ are trend and sinusoidal components correspondingly. It should be noted that in contrast to the standard methods of processing time series presented a method for obtaining IMF starts allocating high-frequency components and completes with a monotonic function or a function with a single peak.

The next stage of the Hilbert-Huang integral transform is application of the Hilbert transform directly to each IMF. This allows one to select from a given IMF two components: the instantaneous magnitude and instantaneous phase. Let us consider this more in details for the real signal $x(t)$. Let us supplement the signal $x(t)$ to an analytic function

$$z(t) = x(t) + ix_H(t), \qquad (14.1.5)$$

where $x_H(t)$ is the Hilbert transform defined as following singular integral

$$x_H(t) = \frac{1}{\pi} P.V. \int_{-\infty}^{+\infty} \frac{x(s)}{t-s} ds. \qquad (14.1.6)$$

P.V. in (14.1.6) stands for Cauchy principal value. Let us rewrite (14.1.5) as follows

$$z(t) = A(t)e^{i\psi(t)}, \qquad (14.1.7)$$

where

$$A(t) = \sqrt{x^2(t) + x_H^2(t)} \qquad (14.1.8)$$

is instantaneous amplitude

$$\psi(t) = \text{arctg} \frac{x_H(t)}{x(t)} \qquad (14.1.9)$$

is phase of the given signal $x(t)$. Then the derivative of (14.1.9) defines the instantaneous frequency

$$\omega(t) = \dot{\psi}(t) = \frac{d}{dt} \text{arctg} \frac{x_H(t)}{x(t)}. \qquad (14.1.10)$$

14.1.4 *Application to Forecasting EPS Parameters*

In order to increase the accuracy of the operation conditions forecasting the "intelligent" neural approach, readers may refer to the papers of [Kurbatsky *et al.* (2011c,d,a)]. Let us now consider the hybrid model for short-term forecast of parameters of expected operating conditions based on the two-stage adaptive neural network approach. The first stage involves decomposition of the time series into intrinsic modal functions and subsequent application of the Hilbert integral transform. At the second stage the computed modal functions and amplitudes are employed as input functions for optimized neural networks block of our approach.

There are three main steps in our model (as shown in Fig. 14.3) as follows:

(1) The EMD algorithm is used to decompose initial non-stationary signal $x(t)$ into several IMFs (see Fig. 14.4). Following the Hilbert transform the corresponding instantaneous amplitude (A_t) and instantaneous frequency are calculated.
(2) The calculated values of IMFs and A_t are used as input values for neural network model.

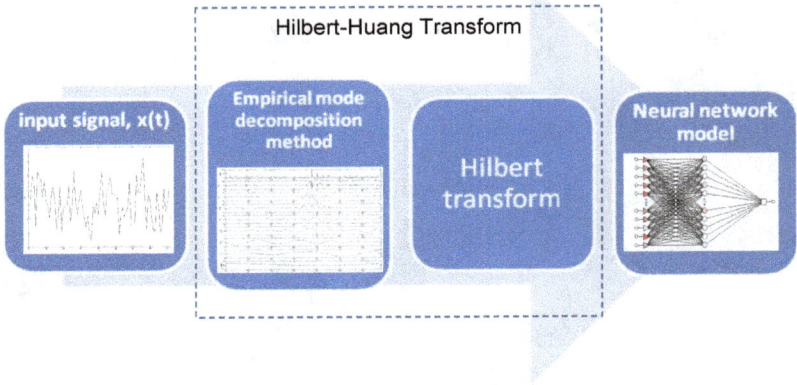

Fig. 14.3.   Optimized hybrid model construction.

(3) The optimization algorithms of neural-genetic selection and sim-
    ulated annealing are used to construct the neural network model.
    This ANN model is learned to forecast the corresponding changes
    of EPS parameters on a given interval of expectation.

Results Analysis   In order to evaluate the accuracy of state variables
the different metrics were employed:
   • The mean absolute percentage error (MAPE):

$$MAPE = \frac{1}{n} \sum_{t=1}^{n} \frac{|x_t - \overline{x}_t|}{x_t} \cdot 100\%. \qquad (14.1.11)$$

   • The mean absolute error (MAE):

$$MAE = \frac{1}{n} \sum_{t=1}^{n} |x_t - \overline{x}_t|. \qquad (14.1.12)$$

   • The root mean squared error (RMSE):

$$RMSE = \sqrt{\frac{1}{n} \sum_{t=1}^{n} (x_t - \overline{x}_t)^2}, \qquad (14.1.13)$$

where A_t is actual value of parameter at t, F_{ts} is forecasted value of parameter at t and n is number of measurements. [Kurbatsky *et al.* (2011e)]

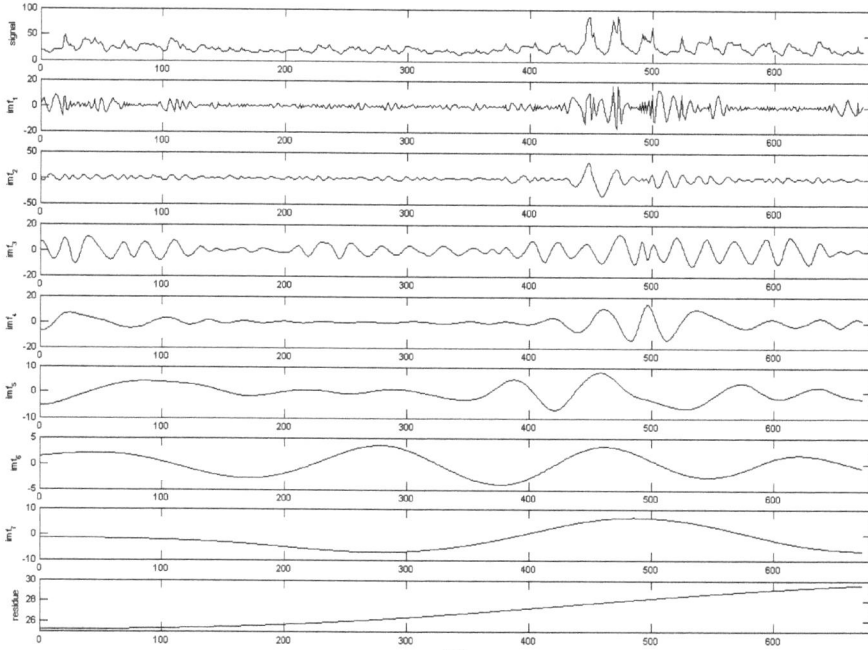

Fig. 14.4.   Set of IMF's $x(t)$.

applied the technique of optimization training ANN for forecasting of voltage magnitude for a lead time interval of 15 min in one of the power systems of Eastern Siberia. The results of the creating of the optimal input sample containing INF's, instantaneous amplitude (A) and instantaneous frequency. Employment the optimization unit (NGIS + SA) has demonstrated the advantage of GRNN architecture (see Table 14.1).

It is to be concluded that optimized ANN training in combination with HHT reduces the average absolute error of 3-5% on the example of short-term forecasting of voltage magnitude (see Fig. 14.4 and Table 14.2).

The importance of the HHT components is the contribution it makes to the success of the model. For a predictive model, success means good forecasting. Often the forecasting relies mainly on a few variables. A good measure of importance reveals those variables. The better the forecasting, the closer the model represents reality, and the more plausible it is that the important variables represent the true cause of forecasting.

Table 14.1   Competition selection algorithm SA in the short-term forecasting voltage magnitude.

Number of SA steps	Type of forecast model	MAE, MW	Number of input neurons	Number of hidden neurons	Performance
1	MLP	7.25	9	8	0.94
2	RBF	7.23	9	11	1.02
3	RBF	7.20	9	8	0.89
4	GRNN	7.13	10	250	1.18
5	GRNN	6.27	10	250	0.88
6	GRNN	6.08	10	250	0.88
7	GRNN	5.98	10	250	0.92
8	GRNN	5.98	10	250	0.88
9	GRNN	5.93	10	250	0.91
10	**GRNN**	**5.91**	**10**	**250**	**0.88**

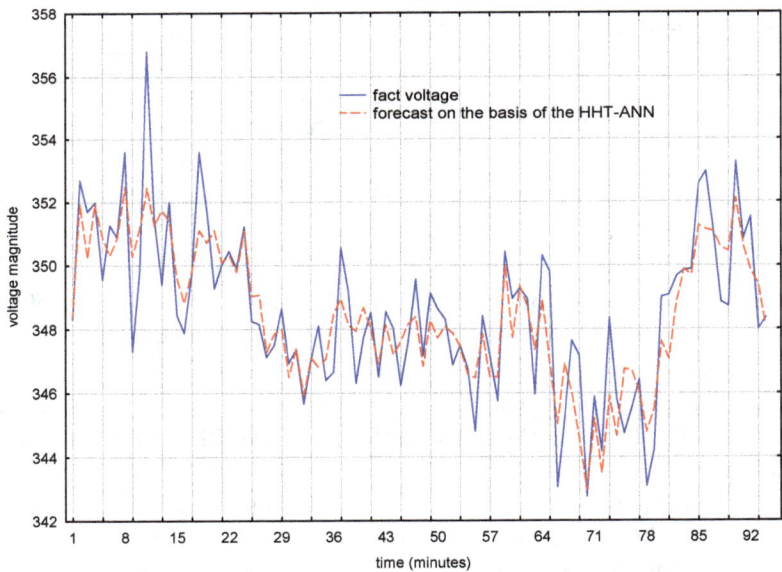

Fig. 14.5.   Forecast of voltage magnitude for a lead time interval of 15 min.

Let us now concentrate on wind speed and direction forecasting as one of the most challenging cases. Here we used the decision tree techniques to rank the importance of variables. These algorithms can output a list

Table 14.2   Comparison of the voltage magnitude forecasts with antici-
pation 15 min for the different models.

Period	Error	Traditional ANN model	Hybrid (HHT-ANN) Model with use of SA method
00:00 – 23:00	MAPE (%)	2.1	0.29
	MAE	2.2	0.99
	RMSE	6.1	1.29

of predictor variables that they believe to be important in predicting the
outcome. If nothing else, researchers can identify a subset of the data to
only include the most "important" variables, and use that with another
model. As it is explained in the paper of [Louppe *et al.* (2013)], every
binary decision tree is separately represented by a tree structure T, from
an input vector $(X_1, \ldots, X_p)$ taking its values in $(\chi_1 * \cdots * \chi_p = \chi)$ to a
output variable $y \in Y$. Every certain node t represents a subset of the
space χ, with the root node being χ itself. Construction of decision trees
usually works top-down, by choosing a variable at each step that splits the
set of items by the binary test $s_t = (X_m < c)$. The internal node t divides
its subset into two subsets corresponding to two children nodes t_L and t_R.
More generally, splits are defined by a (not necessarily binary) partition of
the range χ_m of possible values of a single variable χ_m. For a new instance
the predicted value $\hat{Y}$ is the label of the leaf reached by the instance when it
is propagated through the tree. Algorithms for constructing decision trees
identify at each node t the split $s_t = s^*$ for which the partition of the N_t
node samples into t_L and t_R maximizes the decrease

$$\Delta i(s, t) = i(t) - p_L i(t_L) - p_R i(t_R).$$

With regard to variable importance in Random Forests, [Breiman (2001)]
proposed to add up the weighted impurity (loss) decreases $p(t)i(s_t, t)$ for
all nodes t where X_m is used, averaged over all N_T trees in the forest:

$$Imp(X_m) = \frac{1}{N_t} \sum_T \sum_{t \in T : \nu(s_t) = X_m} p(t)i(s_t, t)$$

and where $p(t)$ is the proportion $\frac{N_t}{N}$ of samples reaching t and $\nu(s_t)$ is a
variable which is used in split s_t. We have examined two decision tree learning techniques: Random Forests and gradient boosting trees. We employ
the decision tree techniques to rank the importance of variables. Also we
use the Gini index as an impurity function.

Valentia wind speed forecasting (Ireland) using the HHT-ANN-SVM model The initial training data is represented by two per-hour time-series: wind speed and wind direction values for 1 year for the Valentia region. Both time series were decomposed into IMFs by the Huang method, and the Hilbert transform was employed to obtain the amplitudes, A and frequency, F. The latter along with these HHT components were used as input values of the selected HHT-ANN and HHT-SVM models. The final 48 hours of the data were reserved for calculating performance statistics.

Figure 14.6 illustrates the feature selection analysis of the modified Valentia wind speed data using regression machine learning algorithms – Boosting Trees and Random Forest. As seen, high-frequency components of the initial wind speed data (IMF1, IMF2, A_1, A_2 and F_1, F_2) have less predictive importance (less than 0.3) by comparison with other components, and therefore they can be excluded from the trained data.

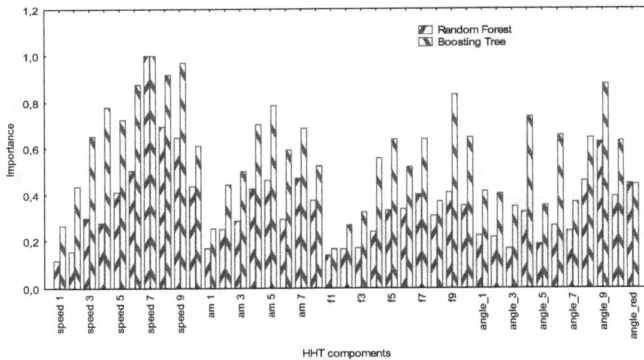

Fig. 14.6. Predictor importance diagram based on Random Forest and Boosting Trees learning algorithms for the Valentia wind speed data.

As a result, the following models are trained and tested: a RBF neural network (HHT-ANN model) and epsilon-SVM regression model (HHT-SVM model). The RBF network is trained by a 2-Phase-RBF scheme (two-step algorithm). After a training-testing process we obtained the following errors: train error – 0.251%, validation error – 0.263% and test error – 0.257%. The obtained RBF network had the following structure: input neurons 25, hidden neurons 162 and output neuron 1. Using a 10-fold cross-validation method the following optimal training constants and RBF kernel parameters were determined: the capacity constant $C = 1$; $\varepsilon = 1.3$;

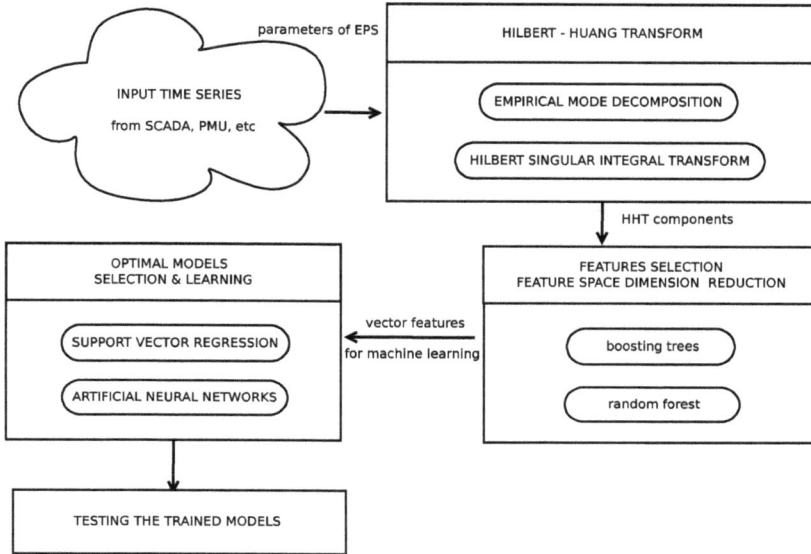

Fig. 14.7.   Data flow diagram of the boosted hybrid forecasting model.

$d = 3; \gamma = 0.03$. The resulting data flow diagram of the boosted hybrid forecasting model os shown in Fig. 14.7.

In this case example a feed forward ANN is tested for performance comparison with the hybrid HHT approach. The candidate input vectors for the FF-ANN-SA model were: the current value of wind speed at time t; previous values of wind speeds at times $(t-1)$, $(t-3)$, $(t-6)$; wind direction at times t and $(t-1)$; and time of day. All input vectors and the target vector were normalized to the range $[0, 1]$, and the data were divided in the ratio 4:1 between training and validation sets, with the final 48 hours reserved for testing as before.

A hybrid simulated annealing/tabu search algorithm for simultaneous optimization of the network structure and weights was applied to the problem of 24h ahead prediction of wind speeds. The structural optimization element of the algorithm uses a network pruning approach, which involves defining an initial, maximal network and then randomly removing intra-neuron connections until a more parsimonious network is arrived at (Fig. 14.8). The inter-neuron connectivity and weights were optimized using the hybrid algorithm prior to subsequent optimization of weights and biases using Levenberg-Marquadt backpropagation.

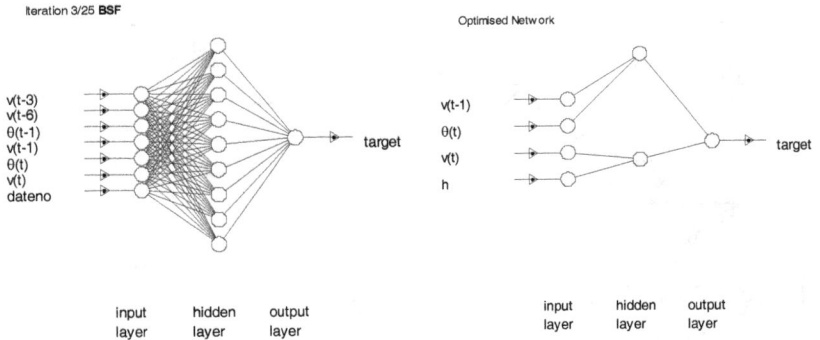

Fig. 14.8.   (a) maximal network; (b) FF-ANN-SA network with optimised structure.

The maximal network was defined as having all seven candidate inputs, and a single hidden layer with nine neurons. Therefore, the maximum number of inter-neuron connections was $N_1 N_2 + N_2 N_3$ where N_1 is the number of inputs, N_2 the number of hidden layer neurons, and N_3 the number of outputs. The cost function specified for the optimization is based on predictive performance and number of inter-neuron connections in the network, with a 3:1 weighting ratio between network complexity and performance. Tan-sigmoidal and linear activation functions were used for the hidden and output layer neurons, respectively. The linear activation function of the output layer was constrained to produce only positive outputs, in order to represent the physical reality of wind speeds which are zero-limited. The algorithm was allowed to run for a maximum of 25 iterations, or until the GL5 early stopping criterion (used in the paper of [Prochelt (1994)]) was satisfied. In Table 14.3 and Figs. 14.9 the testing results for a "24 hour ahead" wind speed forecasting are presented based on the HHT–ANN, HHT–SVM and FF-ANN-SA models. As shown in Table 14.3, the most accurate forecast was given by the hybrid HHT-SVM and HHT-ANN models (MAE for 0-24h period – 0.58 ms and 0.55 ms respectively), compared with the conventional approaches: the FF-ANN-SA model – 2.70 ms. The FF-ANN-SA algorithm did not perform as well as the HHT methods but identified the inputs with the most predictive power as: the current wind speed, the wind speed of the previous hour, the wind direction and the time of day.

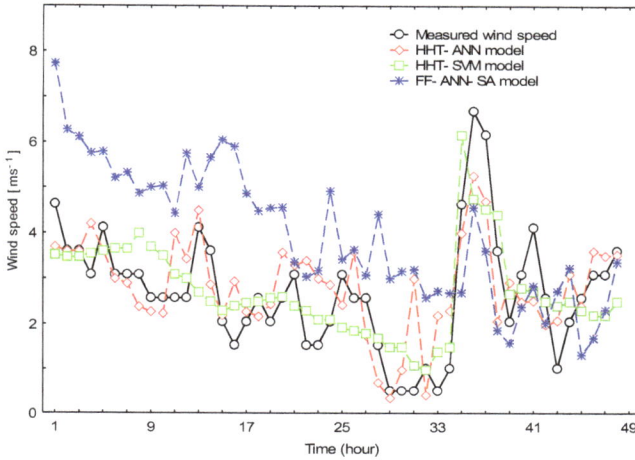

Fig. 14.9. "24-hour ahead" wind speed forecast based on hybrid HHT-ANN and HHT-SVM models.

Table 14.3   Comparison of "24hour ahead" wind speed forecasts on the basis of hybrid and conventional neural approaches for two subsets of the test data.

No.	Test data interval	Error	*HHT-ANN*	*HHT-SVM*	*FF-ANN-SA*
1	0-24h	MAE	0.55	0.58	2.70
		RMSE	0.75	0.88	3.30
2	24-48h	MAE	0.73	0.93	1.54
		RMSE	1.05	1.17	1.69

14.1.5   *Conclusion*

In the modern EPS there is monitoring and forecasting the parameters of the expected operation regime of the EPS, especially it depends on the frequency of the electrical network, the voltage magnitude and the active power flows. An effective way of solving of an optimal input sample formation for ANN trainings is on the basis of use of the nonlinear optimization algorithms, namely SA method and NGIS, which provides a procedure for selecting the best predictive model for each sample. We apply the technique of optimization training ANN for short-term forecasting state variable of the Eastern Siberia EPS and Valentia wind speed forecasting in Ireland using of the combinations of machine learning technologies and HHT. Results of application of the HHT-ANN and HHT-SVM models for the electricity prices forecasting are presented in the articles of [Kurbatsky *et al.* (2011b)] and [Kurbatsky *et al.* (2014)]. The retrospective data from Australian electricity market (AEMO Market Operator) were employed as benchmark data.

As footnote let us conclude that optimized ANN and SVM models in combination with HHT reduces the average absolute error of 3-5% on the examples of short-term forecasting of voltage magnitude, electricity prices and wind speed and direction.

14.2   Non-stationary Autoregressive Model for On-line Detection of Inter-area Oscillations in Power Systems

This chapter addresses early on-line detection of inter-area electro-mechanical oscillations in power systems using dynamic data such as currents, voltages and angle differences measured across transmission lines in real time. The main objective is to give the transmission operator qualitative information regarding stability margins. Below we discuss the approach where the observed signal is modeled with a non-stationary second order autoregressive model. Bayesian estimation of the system is based on the forgetting approach. The stability margins are obtained as posterior probabilities that the poles of the estimated system are unstable. The approach is demonstrated on real retrospective data recorded in a 500 kV power grid and voltage data obtained by numerical simulations.

The results presented in the section were obtained within Russian-Czech Inter-Academia project No. 67 in collaboration with V. Šmídl, Yu. A. Grishin and D. Panasetsky. Readers may also refer to the publications of [Sidorov *et al.* (2010a,b)].

14.2.1   *Introduction*

Power systems are characterized by many modes of electro-mechanical oscillations caused by interactions among its components. During such oscillations, mechanical kinetic energy is exchanged between synchronous generators as electric power flows through the network. For example, one generator rotor could swing relative to another. The inter-area modes are usually associated with groups of machines swinging relative to other groups across a relatively weak transmission line. These impacts cause oscillations in state variables of the electric system such as voltage, current, power and frequency, which are conventionally measured by PMU devices (Phasor Measurement Units).

The amplitudes of the swinging state variables are mainly determined by the following factors:

- The position of the subsystem in the whole power system.
- The distribution of the natural damping elements such as series resistance of the lines, and shunt resistance of the loads.
- The number and position of special damping controllers, e.g., Power System Stabilizer (PSS) and controllers of different FACTS devices (SVC, TCSC, etc).

PSS controllers are among the most effective and robust solutions as reported by [Messina *et al.* (2001)], SVCs are also widely used for stabilization by modulating voltage at strategic locations of the power system (see the example of Mexico System in publications of [Messina *et al.* (2001); Banejad (2004)]).

There are two distinct types of problematic oscillations in power systems: local mode oscillations and inter-area oscillations. Local mode oscillations occur when a generator (or group of generators) under voltage regulator control at a station is oscillating against the rest of the system. Inter-area oscillations involve combinations of many machines on one part of a system swinging against machines on another part of the system. It is to be noted that the local mode of oscillations are well damped by the traditional PSS controllers, but normally fails the inter-area ones as shown by [Zhijian *et al.* (2008)].

In this chapter, we address the problem of robust detection of inter-area oscillations. This problem needs more sophisticated approaches in order to ensure accurate monitoring of system dynamics and reliable detection of dangerous oscillations with noise-polluted PMU measurements. Note that oscillations themselves are not necessarily dangerous as long as they do not become unstable, here readers may refer to results of [Korba (2007)]. The key objective of this chapter is to demonstrate the Volterra type non-statinary models in design an algorithm evaluating stability margins.

14.2.2 *Chapter Summary*

The chapter is organized as follows. In Subsection 14.2.3 we discuss the unifying view on the Volterra theory and polynomial modelling as a kernel regression. In Subsection 14.2.4, we provide the motivation of our studies, next we briefly discuss the related techniques employed for on-line oscillations prediction. In Subsection 14.2.5, we present oscillations detection algorithm based on regularized exponential forgetting suitable for non-stationary data analysis of power systems. In Subsection 14.2.6, we apply the designed technique for real retrospective data corresponding to inter-area oscillations event recorded in 500kV power grid and for the data obtained by numerical simulations. The chapter is concluded with Subsection 14.2.7 which contains the remarks and possible ways for improvements of the proposed techniques.

14.2.3   *Volterra Theory and Polynomial Regression*

For practical applications we consider a nonlinear, discrete-time and time-invariant input-output relationship $y_t = f[x_t, x_{t-1}, \cdots, x_1]$, where x_t is observed input signal and y_t denote the output samples at time moment t, $t = \overline{1, N}$. Such nonlinear mapping could have infinity memory. But in practice the following finite memory truncated relationships are conventionally employed $y_t = f[x_t, x_{t-1}, \cdots, x_{t-L+1}]$ with finite L. Under smoothness conditions, this input-output relationship can be approximated by the finite Volterra expansion truncated to a finite-order P as

$$y_t = \sum_{p=0}^{P} H_p[x_t, x_{t-1}, \cdots, x_{t-L+1}] + v_t,$$

where v_t stands for unmodeled dynamics and measurement (observation) noise, which is conventionally assumed to be zero-mean and independent of $x_t, x_{t-1}, \cdots, x_{t-L+1}$. Here H_p denotes the output of the pth order term of truncated Volterra series of transfer function $K_p[s_1, \cdots, s_p]$ which is presented (here readers may refer to the book of [Mathews and Sicuranza (2000)]) as follows

$$H_p[x_t, x_{t-1}, \cdots, x_{t-L+1}] \equiv \sum_{s_1=0}^{L-1} \cdots \sum_{s_p=0}^{L-1} K_p[s_1, \cdots, s_p] \prod_{i=1}^{p} x_{n-s_i}. \quad (14.2.1)$$

For continuous-time analogue, which has been the subject of extensive study in previous Part 2 (Nonlinear Models Singularities and Control) of this book, we studied the following Volterra model

$$y(t) = H_P(x(t)) \equiv \sum_{p=1}^{P} \int_0^t \cdots \int_0^t K_p(s_1, \ldots, s_p) \prod_{i=1}^{p} x(t - s_i) \, ds_i.$$

It is based on the analogous of the famous Weierstrass approximation theorem. Here we refer to the classical Fréchet theorem [Fréchet (1910)] and it's following generalization. According to the paper of [Baesler and Daugavet (1990)] for any continuous casual (Volterra) initial $(F(x(0)) = 0)$ mapping $y(t) = F(x(t))$, $F : \mathcal{K} \to C_{[0,T]}$, $\mathcal{K}$ is compact in $C_{[0,T]}$, and any $\varepsilon > 0$ there is N such that for all $x(t) \in K$ the following equality holds $\|F(x(t)) - P_N(x(t))\|_{C_{[0,T]}} < \varepsilon$. Readers may also refer to the paper of [Makarov and Khlobystov (1997)] for the theoretical results regarding polynomial interpolation of operators.

The objective is to estimate the Volterra kernels (transfer functions) $K_p[s_1, \cdots, s_p]$ for $p = \overline{0, P}$ and $s_i = \overline{1, L-1}$, given the input-output samples $\{x_t, y_t\}_{t=1}^N$ and upper bounds on the expansion order P and the memory size L. It's well known that the main drawback of such Volterra models is their parametric complexity implying the need to estimate a huge number of parameters. For more details regarding the comprehensive analysis of the state-of-the-art Volterra series modifications and methods for the parametric complexity reduction readers may refer to the paper by[Favier *et al.* (2012)]. The adaptation of the Volterra series models for sparse data is presented in the work by [Kekatos and Giannakis (2011)]. The model predictive control (MPC) algorithms based on autoregressive and non-autoregressive Volterra model are considered in the paper of [Gruber *et al.* (2011)]. It is to be noted that MPC is one of the most popular advanced control technique in industry due to the intuitive control problem formulation and because of the ability to deal with economic objectives and operating constraints.

The polynomial kernel regression generalizes the conventional Volterra expansion for MISO dynamical systems and consists in approximation a nonlinear function $y_t = f(\{x_{t,l}\}_{l=0}^{L-1})$ for the case of the input vector $x_{t,0}, \cdots, x_{t,L-1}$, and t is not necessarily a time index. If we select $x_{t,l}$ as x_{t-l} for $l = \overline{0, L-1}$ we observe the Volterra model is a special case of the polynomial kernel regression, here readers may refer to the book of [Mathews and Sicuranza (2000)] and papers of [Franz and Schölkopf (2006); Kekatos and Giannakis (2011)]. [Franz and Schölkopf (2006)] outlined that all the properties of discrete Volterra integral models theory are preserved by using the regularized polynomial kernel regression and the estimation process complexity is independent of the order of nonlinearity.

In this section we consider the application of non-stationary autoregressive model of on-line detection of inter-area oscillations in power systems.

14.2.4 *Problem Statement*

State-of-the-art Techniques The deregularization of the power market has caused substantial demand for the development of new tools for electro-mechanical oscillations prediction. Many classical non-adaptive algorithms such as Yule-Walker, Burgs, lattice and Prony's methods (see papers of [Hauer (1991)], [Ledwich and Palmer (2001)] and [Hemmingsson *et al.* (2001)]) have been applied in the field. Also recursive least squares (RLS) and least squares (LMS) methods are typical solutions. Recently, Kalman filtering (KF) techniques has been employed by [Korba *et al.* (2003)] for on-line oscillations prediction.

Techniques Comparison These methods are typically based on treatment of the underlying process as a linear system. Detection of the oscillations is based on two basic results of linear systems theory:

(1) poles of oscillating linear systems have non-zero imaginary components,
(2) poles of unstable linear systems have absolute values greater than 1.

We use these facts to detect oscillations as follows: (i) the observed process is locally approximated by a linear system, (ii) parameters of the linear system are estimated, (iii) poles of the system are computed from the estimates, and (iv) stability and oscillatory behavior is analyzed.

The methods mentioned above differ typically in (i) and (ii). For example, detection methods based on fixed windows assume that all data in the window were generated by the estimated linear system with identical weight. An alternative is represented by RLS with discounting which assumes exponential decrease of importance of older data records. The difference in (ii) is typically in the assumption whether the variance of the measurement noise is known (Kalman filter) or unknown (RLS-type methods). However, the methods rarely differ in (iii) and (iv) where a point estimate of the poles is being analyzed. We aim to address this issue.

The traditional point estimate approach provides a single option for all poles for given time without any uncertainty bounds on the result. Thus, it is hard to assess the reliability of this value. In present section, we summarize the results of [Sidorov *et al.* (2010a,b)] concerned with the Bayesian approach, i.e., we develop full posterior density of the parameters of the linear system and transform this density to density on poles. For details regarding treatment of the variational Bayes approximation in signal processing readers may refer to the book of [Šmídl and Quinn (2006)].

14.2.5 *Oscillations Detection Algorithm*

We represent the signal as a second-order linear system with unknown time-variant parameters:

$$y_t = a_t y_{t-1} + b_t y_{t-2} + c_t + \sigma_t e_t, \qquad (14.2.2)$$

where y_t is the observed signal, a_t, b_t, c_t, σ_t are its unknown parameters, and e_t is Gaussian noise with zero mean and unit variance, $e_t = \mathcal{N}(0, 1)$.

In the probabilistic formulation, (14.2.2) defines probability density function (pdf) of the observed random variable y_t:

$$p(y_t | y_{t-1}, y_{t-2}, a_t, b_t, \sigma_t) = \mathcal{N}(a_t y_{t-1} + b_t y_{t-2} + c_t, \sigma^2). \qquad (14.2.3)$$

Estimation of system (14.2.3) with stationary parameters is a well known task in statistics, with posterior density of Normal-inverse-Gamma type. Extension of this approach to a non-stationary system can be achieved by specification of the parameter evolution model, $p(a_t, b_t, c_t, \sigma_t | a_{t-1}, b_{t-1}, c_{t-1}, \sigma_{t-1})$. The specific choice of such a model yields a Bayesian filtering task, which can be solved in some cases by the Kalman filter .

However, we consider a simpler alternative, known as forgetting (or discounting). In this approach, the time-variant system is treated similarly to the time-invariant system, but the resulting sufficient statistics is multiplied by a constant ϕ, $0 < \phi < 1$. In effect, the delayed data records, y_{t-k}, are weighted by ϕ^k, which is equivalent to the application of an exponential weighting. However, this simple approach has some shortcomings, such as numerical instability when the data are not informative.

We will apply an improved version of forgetting suggested by [Kulhavý and Kraus (1996)], where regularized exponential forgetting is formalized as follows:

$$
\begin{aligned}
p(a_t, b_t, c_t, \sigma_t | y_1, \ldots, y_t) &\propto p(y_t | y_{t-1}, y_{t-2}, a_t, b_t, c_t, \sigma_t) \\
&\times p(a_{t-1}, b_{t-1}, c_{t-1}, \sigma_{t-1} | y_1, \ldots, y_{t-1})^{\phi} \\
&\times \bar{p}(a_{t-1}, b_{t-1}, c_{t-1}, \sigma_{t-1} | y_1, \ldots, y_{t-1})^{1-\phi}.
\end{aligned}
$$

$$(14.2.4)$$

Here, $\bar{p}(\cdot)$ denotes an *alternative* probability of the parameters. This probability expresses an alternative (prior) knowledge about location of the parameters.

Posterior Density   One advantage of (14.2.4) is that for system (14.2.2) it preserves posterior density of the Normal-inverse-Gamma type,

$$p(a_t, b_t, \sigma_t) = \mathcal{N}i\mathcal{G}(V_t, \nu_t), \tag{14.2.5}$$

the statistics of which are recursively computed as follows:

$$V_t = \phi V_{t-1} + [y_t, y_{t-1}, y_{t-2}, 1]'[y_t, y_{t-1}, y_{t-2}, 1] + (1 - \phi)\bar{V}, \tag{14.2.6}$$

and $\nu_t = \phi \nu_{t-1} + 1 + (1 - \phi)\bar{\nu}$. Here, $\bar{V}, \bar{\nu}$ denote statistics of the alternative pdf.

Important moments of this posterior density are mean value,

$$\left[\hat{a}_t, \hat{b}_t, \hat{c}_t\right]' = [V_{2,1}, V_{3,1}, V_{4,1}]\, C, \tag{14.2.7}$$

$$C = \begin{bmatrix} V_{2,2} & V_{2,3} & V_{2,4} \\ V_{3,2} & V_{3,3} & V_{3,4} \\ V_{4,2} & V_{4,3} & V_{4,4} \end{bmatrix}^{-1}$$

which is equivalent to the result of RLS with discounting (under the choice of $\bar{V} = 0$). Covariance of the autoregressive parameters is:

$$\mathrm{cov}([a, b, c]) = \frac{V_{1,1}}{\nu_t - 7} C. \tag{14.2.8}$$

Since the main parameters of interest are parameters $[a_t, b_t]$ we marginalize (14.2.5) to obtain marginal density of Student-t type. An important property of this density is that it is not as sharply concentrated as a Gaussian, since it assigns higher probability to values distant from the mean. The difference is greatest for $\nu_t < 20$, which arise for $\phi < 0.95$. Since we usually choose $\phi > 0.95$ we consider (14.2.7) and (14.2.8) to be an adequate approximation. For more details, see the book of [Kárný *et al.* (2005)].

Prior Distribution   The prior distribution should reflect information that is not available in the processed data. Typically, information about possible ranges of parameters, their expected ratios, etc. The same is true about the alternative density $\bar{p}(\cdot)$ in (14.2.4). Note that non-informative data $y_t = y_{t-1}, \forall t$ can be explained by infinitely many combinations of parameters a_t, b_t, c_t. In standard RLS, this degeneracy would manifest itself by the loss of rank of matrix (14.2.6) yielding numerically unstable estimates (14.2.7). The stabilized forgetting avoids this danger via positive definite matrix $\overline{V}_t$. In case of non-informative data, the estimates are fully determined by this matrix.

In all our simulations we made the following choice:

$$\overline{V}_t = \mathrm{diag}([1e-2, 1e-3, 1e-3, 1e-5]).$$

Under this choice, the expected values of parameters are $[\hat{a}_t, \hat{b}_t, \hat{c}_t, \hat{\sigma}_t] = [0, 0, 0, 0.0033] \pm [3, 3, 36, 0.0067]$. This encodes the following information:

- the linear coefficients, a_t and b_t, are assumed to be closer to zero that the estimated constant c_t
- the variance of observations σ_t is small.

The influence of this alternative distribution on the posterior (14.2.6) is negligible when the data are informative.

Oscillations Detection The probability of unstable oscillation is computed as the probability of unstable oscillating poles of the system. The poles of the system (14.2.2) are:

$$p_{1,2} = \frac{a_t \pm \sqrt{a_t^2 + 4b_t}}{2}.$$

The system is oscillating when the poles are imaginary (i.e., $a_t^2 < -4b_t$) and the system is unstable when $|p_{1,2}| > 1$, i.e.,

$$|p_{1,2}| = \left| \left(\frac{a_t}{2}\right)^2 - \frac{a_t^2 + 4b_t}{4} \right| = |b_t| > 1.$$

These two inequalities define a space over which we need to integrate the posterior density (14.2.5). This demanding task was addressed via Monte Carlo sampling in the paper of [Sidorov *et al.* (2010b)].

In this chapter, we propose to use an approximation motivated by the fact that we are most often interested in detecting oscillations around the stability boundary, $b_t = -1$. Around this boundary, the first inequality is approximated by $|a_t| < 2$. Moreover, we expect the oscillation to be slower than the sampling frequency, i.e., $a_t > 0$ we can only test probability of $a_t < 2$. Under this approximation and approximation of the posterior by a Gaussian density, the probability of unstable oscillation is computed as

$$\Pr(unstab.oscil.) = \Pr(a_t < 2)\Pr(b_t < -1)$$

$$= \frac{1}{2}\left(1 - \mathrm{erf}\frac{\hat{a}_t - 2}{\sqrt{2}\mathrm{var}(a_t)}\right)\frac{1}{2}\left(1 - \mathrm{erf}\frac{\hat{b}_t + 1}{\sqrt{2}\mathrm{var}(b_t)}\right)$$

$$(14.2.9)$$

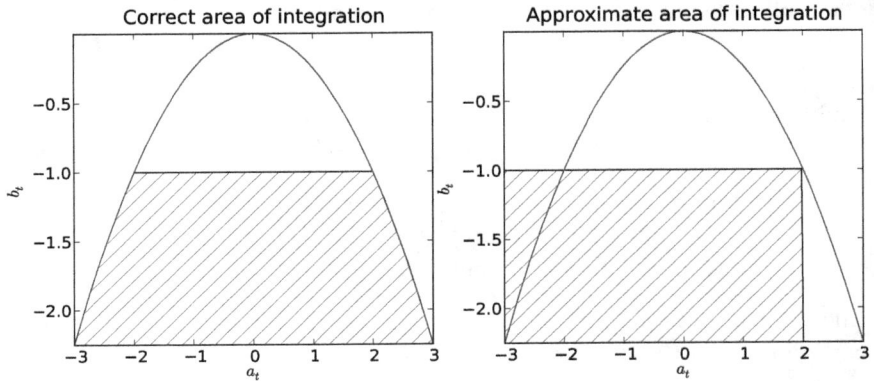

Fig. 14.10.   Illustration of integration area for evaluation of oscillation risk.

where erf is the error function. This approximation tends to underestimate the risk of oscillations since the approximate area of integration is smaller than the correct one, see Fig. 14.10. Thus, it can serve as a lower bound of the true risk. On the other hand, probability of instability

$$\Pr(instability) = \Pr(b_t < -1)$$

$$= \frac{1}{2}\left(1 - \text{erf}\frac{\hat{b}_t + 1}{\sqrt{2}\text{var}(b_t)}\right) \qquad (14.2.10)$$

may serve as an upper bound.

Final oscillation detection algorithm is then as follows:

Off-line: choose initial alternative statistics, $\bar{V}, \bar{\nu}$ and forgetting factor ϕ.

On-line: at each time t do:

(1) update statistics V, ν using (14.2.6),
(2) compute posterior mean and variance via (14.2.7) and (14.2.8)
(3) compute probability of unstable oscillations using (14.2.9) or (14.2.10)

14.2.6   *Results Analysis*

In order to verify the designed techniques, two different sets of data are considered. The first data set is from real PMU measurements. The second data set is obtained by means of numerical simulations.

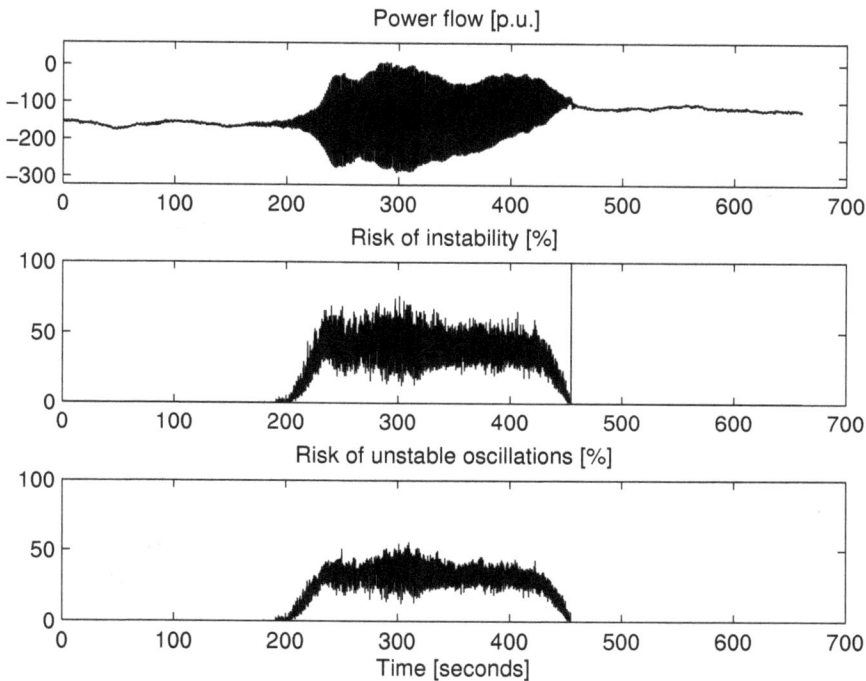

Fig. 14.11. Analysis of real data with $\phi = 0.97$. **Top**: observed power flow through 500kV transmission line. **Middle**: Risk of instability (14.2.10) **Bottom**: Risk of unstable oscillations. (14.2.9).

Real Data Analysis The first case study represents analysis of a real retrospective time series. In particular, time series of power flow (see Fig. 14.11) has been registered using PMU devices from 500 kV power grid (as reported by [Wu (2006)]) with sampling period of 20 ms. Note that the condition of instability (14.2.10) allows to detect instability of the system, however, the condition of unstable oscillation (14.2.9) provides smoother estimates without the outlier artifact at the end of the instability region.

Model Data Analysis The second case study represents analysis of data obtained by means of numerical simulations. A classical model of the two-area system shown in Fig. 14.12 was used for inter-area oscillations modelling.

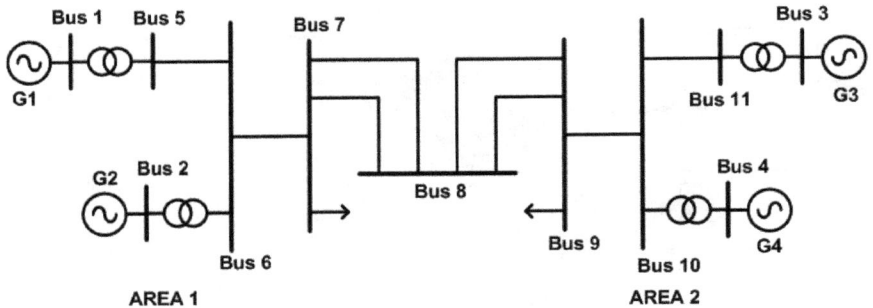

Fig. 14.12.   An 11-bus test system.

Test system parameters are represented in the book of [Kundur (1994)]. Area 1 and Area 2 are sending and receiving subsystems accordingly. White noise having signal-to-noise ratio 40db and spectral width 0.5 Hz (accidental load variation) was added to the load which makes the simulations more complicated and realistic. The sampling period is 0.02 second. During the simulation, the following scenario of changing an active power flow from Area 1 to Area 2 has been studied:

- From 50 to 115 second, increase of active power output of G2 from 500 MW to 660 MW. Active power increase rate 2.5 MW per second;
- From 165 to 230 second, decrease of active power output of G2 from 660 MW to 500 MW. Active power decrease rate 2.5 MW per second.

Variation of voltage on Bus 8 during the simulation process and corresponding risk of instability are displayed in Fig. 14.13.

Nowadays, PMU measurement technologies provide a good observability of real-time processes in power systems. It is to be noted that when having started, the inter-area oscillations can be easily detected by a system operator. However, correctly assessing where they actually start is a non-trivial task. The computed probability of both real and modeled oscillations marks the start of the oscillations quite clearly. Thus, after detailed studies it is possible that the considered method can be adopted to give the transmission operator qualitative information regarding stability margins.

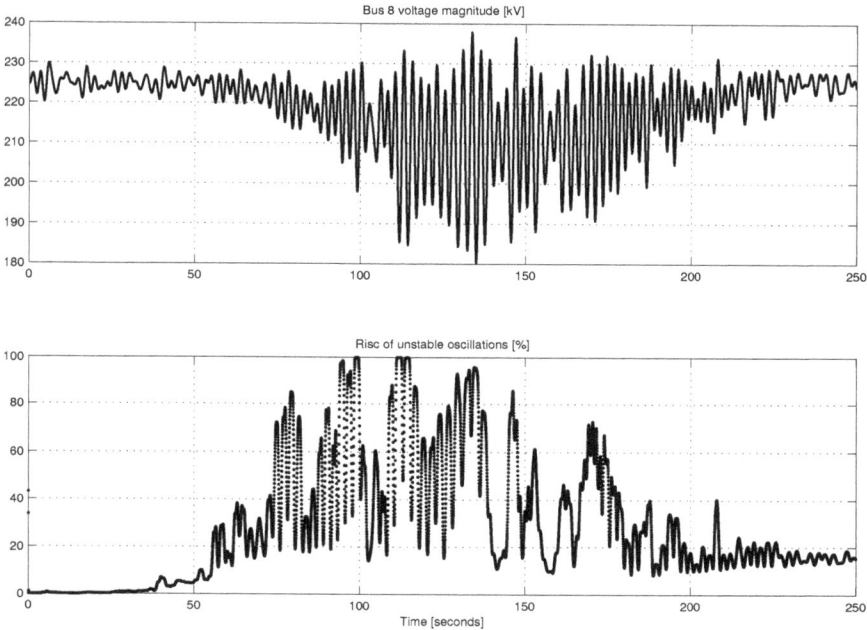

Fig. 14.13.   Numerical simulation. Results for voltage amplitude data. **Top**: observed data. **Bottom**: computed probability of unstable oscillations.

14.2.7   *Conclusion*

We addressed the problem of robust detection of inter-area electro-mechanical unstable oscillations in power systems using dynamic data registered by PMU. Oscillating systems have been modeled by a second order autoregressive system. The stability is assessed by probability of unstable poles of the estimated approximate linear system.

The algorithm was tested on recorded retrospective data, in particular power flow through 500kV transmission line was employed. We employed the data of the event reported by [Wu (2006)]. We found that the computed probability of oscillations clearly marks the start of the unstable oscillations. We conjecture that the designed algorithm has an advantage over the conventional approach based on Kalman filtering, because it is able to estimate covariance of the observation σ_t. For deeper understanding of its properties, the algorithm has been tested on voltage data obtained by numerical simulations of well studied and understood scenarios.

Bibliography

Abdallah, N. B., Degond, P., and Mehats, F. (1997). Mathematical Models of Magnetic Insulation, in *Rapport Interne, No. 97.20 MIP (Université Poul Sabatier* (Toulouse, France), pp. 1–38.

Aiordachioaie, D., Ceanga, E., Keyser, R. D., and Naka, Y. (2001). Detection and Classification of Nonlinearities based on Volterra Kernels Processing, *Engineering Applications of Artificial Intelligence* **14**, pp. 497–503.

Aizenberg, I. and Butakoff, C. (2002). Frequency Domain Median-like Filter for Periodic and Quasi-Periodic Noise Removal, in *Proc. SPIE: Image Processing: Algorithms and Systems 2002* (San Jose, USA), pp. 46—57.

Aizenberg, I., Butakoff, C., Astola, J. T., and Egiazarian, K. (2002). Nonlinear Frequency Domain Filter for Quasi-Periodic Noise Removal, in *Spectral Methods and Multirate Signal Processing (SMMSP'2002), Proc. of TICSP Workshop* (Toulouse, France), pp. 147—153.

Amidror, I. (2000). *The Theory of the Moiré Phenomena* (Kluwer Academic Publ.).

Anselone, P. M. (1967). Uniform Approximation Theory for Integral Equations with Discontinuous Kernels, *SIAM J. Numer. Anal.* **4**, pp. 245–253.

Apartsyn, A. S. (2003). *Nonclassical Linear Volterra Equations of the First Kind* (Walter de Gruyter).

Apartsyn, A. S. (2013). Polynomial Volterra Integral Equations of the First Kind and the Lambert Function, *Proceedings of the Steklov Institute of Mathematics* **280**, 1, pp. 26–38.

Apartsyn, A. S. and Solodusha, S. V. (2004). Test Signal Amplitude Optimization for Identification of the Volterra Kernels, *Automation and Remote Control* **3**, pp. 116–124.

Apartsyn, A. S., Solodusha, S. V., and Spiryaev, V. A. (2013). Modeling of Nonlinear Dynamic Systems with Volterra Polynomials: Elements of Theory and Applications, *International Journal of Energy Optimization and Engineering* **2(4)**, pp. 16–43.

Arthur, R. and Weeks, J. (1996). *Fundamentals of Electronic Image Processing* (IEEE Press).

Asanov, A. (1992). *Regularization, Uniqueness and Existence of Solutions of*

Volterra Equations of the First kind (Utrecht: VSP Publ.).

Banejad, M. (2004). *Identification of Damping Contribution from Power System Controllers* (Queensland University of Technology. PhD Thesis).

Barbashin, E. A. (1970). *Introduction to the Theory of Stability* (Wolters-Noordhoff (Trans. from Russian)).

Belbas, S. A. and Bulka, Y. (2010). Numerical Solution of Multiple Nonlinear Volterra Integral Equations, *Applied Mathematics and Computation* **9**, 217, pp. 4791–4804.

Bel'tukov, B. A. and Shil'ko, G. S. (1973). The Extrapolation Method for Solving a Nonlinear Integral Equation in the Nonregular Case, *Sb. Vychisl. Mat., Irkutsk Ped. Inst., Irkutsk*, **1**, pp. 104–119.

Bendat, J. S. (1990). *Non-Linear System Analysis and Identification from Random Data* (Wiley, New York).

Boikov, I. V. and Tynda, A. N. (2003). Approximate Solution of Nonlinear Integral Equations of the Theory of Developing Systems, *Differential Equations* **9**, 39, pp. 1277–1288.

Breiman, L. (2001). Random Forests, *Machine Learning* **45**, pp. 5—32.

Brunner, H. (1997). 1896–1996: One Hundred Years of Volterra Integral Equations of the First Kind, *Applied Numerical Mathematics* **24**, pp. 83–93.

Brunner, H. (2004). The Numerical Analysis of Functional Integral and Integrodifferential Equations of Volterra Type, *Acta Numerica* **13**, pp. 55–145.

Brunner, H. and Houwen, P. J. (1986). *The Numerical Solution of Volterra Equations* (CWI, North-Holland, Amsterdam).

Bryuno, A. D. (2000). *The Power Geometry in Algebraic and Differential Equations* (Elsevier, North-Holland).

Buffoni, B. and Toland, J. (2003). *Analytic Theory of Global Bifurcation* (Princeton University Press).

Cho, N. I., Choi, C.-H., and Lee, S. U. (1989). Adaptive Line Enchancement by Using an IIR Lattice Notch Filter, *IEEE Trans. on Acust., Speech, Signal Processing* **37**, 4, pp. 585—589.

Chua, L. O. and Liao, Y. (1991). Measuring Volterra Kernels (iii): How to Estimate the Highest Significant Order, *Int. J. Circuit Theory Appl.* **19**, 3, pp. 189–209.

Chua, L. O. and Tang, Y. S. (1982). Nonlinear Oscillations via Volterra Series, *IEEE Trans. Circuits and Systems* **3**, 29, pp. 150–168.

Coddington, E. A. and Levinson, N. (1955). *Theory of Ordinary Differential Equations* (McGraw Hill, NY).

Corduneanu, C. (2002). *Functional Equations with Causal Operators* (Taylor and Francis).

Denisov, A. M. and Lorenzi, A. (1995). On a Special Volterra Integral Equation of the First Kind, *Boll. Un. Mat. Ital. B.* **7**, 9, pp. 443–457.

Diouf, C., Telescu, M., Cloastre, P., and Tanguy, N. (2012). On the Use of Equality Constraints in the Identification of Volterra–Laguerre Models, *IEEE Signal Processing Letters* **12**, 19, pp. 857–860.

Dmitriev, V. I. and Zaharova, E. V. (1968). Numerical Solution to Some Fredholm Integral Equations of the First Kind, *Numerical Methods and Programming*.

MSU Publ. **10**, pp. 49–54.

Dolezal, V. (1967). *Dynamics of Linear Systems* (Academia, Prague).

Doyle, F. J., Ogunnaike, B. A., and Pearson, R. K. (1995). Nonlinear Model-Based Control using Second-Order Volterra Models, *Automatica* **31**, pp. 697–714.

Elsgoltz, L. E. (1964). *Qualitative Methods in Mathematical Analysis* (American Math. Soc.).

Espiritu, J. F. and Coit, D. W. (2008). A Component Replacement Model for Electricity Distribution Systems, *The Engineering Economist* **53**, 4, pp. 318–339.

Evans, C., Rees, D., Jones, L., and Weiss, M. (1995). Probing Signals for Measuring Nonlinear Volterra Kernels, in *Proc. Instrumentation and Measurement Technology Conf. (IMTC-95), Boston, MA* (IEEE, Washington, D.C., IFAC), pp. 10–15.

Evans, G. C. (1910). Integral Equation of the Second Kind with Discontinuous Kernel, *Transactions of the American Mathematical Society* **11**, 4, pp. 393–413.

Falaleev, M. V. (2000). Fundamental Operator-functions of Singular Differential Operators in Banach Spaces, *Siberian Mathematical Journal* **41**, 5, pp. 960–973.

Favier, G., Kibangou, A. Y., and Bouilloc, T. (2012). Nonlinear System Modeling and Identification using Volterra-PARAFAC Models, *Int. J. Adapt. Control Signal Process* **26**, pp. 30–53.

Franz, M. and Schölkopf, B. (2006). A Unifying View of Wiener and Volterra Theory and Polynomial Kernel Regression, *Neural Comput.* **18**, pp. 3097–3118.

Fréchet, M. (1910). Sur les Fonctionnelles Continues, *Annales scientifiques de l'École Normale Supérieure* **27**, pp. 193–216.

Gamm, A. Z., Glazunova, A. M., Grishin, Y. A., Kurbatsky, V. G., Sidorov, D. A., Spiryaev, V. A., and Tomin, N. V. (2011). Methods for Forecasting of State Variables of Electric Power Systems for Monitoring and Control, *Electrichestvo* **5**, pp. 12—20.

Glushkov, V. M., Ivanov, V. V., and Yanenko, V. M. (1983). *Modeling of Evolving Systems* (Moscow: Nauka).

Grebennikov, E. A. and Ryabov, Y. A. (1979). *Constructive Methods in the Analysis of Nonlinear Systems* (Nauka. Moscow).

Gruber, J. K., Bordons, C., Bars, R., and Haber, R. (2011). Nonlinear Predictive Control of Smooth Nonlinear Systems Based on Volterra Models. Application to a Pilot Plant, *International Journal of Robust and Nonlinear Control* **20**, pp. 1817–1835.

Haikin, S. (2009). *Neural Networks and Learning Machines* (Prentice Hall).

Halmos, P. R. (1978). *Measure Theory (Graduate Texts in Mathematics)* (Springer, New York).

Hauer, J. (1991). Application of Prony Analysis to the Determination of Modal Content and Equivalent Models for Measured Power System Response, *IEEE Trans. on Power Systems* **6**, 3, pp. 1062–1068.

Hemmingsson, M., Samuelsson, O., Pedersen, K., and Nielsen, A. (2001). Estimation Electro-Mechanical Parameters using Frequency Measurements, in *Power Engineering Society Winter Meeting, 2001*, Vol. 3 (IEEE, Washington, D.C.), pp. 1172–1177.

Hentschel, C. (2001). Video Moiré Cancellation Filter for High-Resolution crts, *IEEE Transactions on Consumer Electronics* **47**, 1, pp. 16—24.

Hinamoto, T., Ikeda, N., Nishimura, S., and Doi, A. (2000). Design of Two-Dimentional Adaptive Digital Notch Filters, in *5th International Conference on Signal Processing (WCCC-ICSP 2000)*, pp. 538—542.

Hong, W. C. (2013). *Intelligent Energy Demand Forecasting*, Lecture Notes in Energy (Springer London).

Hritonenko, N. and Yatsenko, Y. (1996). *Modeling and Optimization of the Lifetime of Technologies* (Kluwer Academic Publushers).

Hritonenko, N. and Yatsenko, Y. (1999). *Mathematical Modeling in Economics, Ecology and the Environment* (Kluwer Academic Publushers).

Hritonenko, N. and Yatsenko, Y. (2003). *Applied Mathematical Modelling of Engineering Problems* (Kluwer Academic Publushers).

Hritonenko, N. and Yatsenko, Y. (2005). Turnpike and Optimal Trajectories in Integral Dynamic Models with Endogenous Delay, *Automations and Remote Control* **127**, pp. 109–127.

Huang, N. E., Zheng, S., Long, R. L. S. R., and et al. (1998). The Empirical Mode Decomposition and the Hilbert Spectrum for Non-linear and Non-stationary Time Series Analysis, *Proc. Royal Soc. London, Ser. A: Math., Phys. and Engineer. Sci.* **454**, 1971, pp. 903—995.

Ivanov, D. V., Karaulova, I. V., Markova, E. V., Trufanov, V. V., and Khamisov, O. V. (2006). Control and Power Grid Development: Numerical Solutions, *Automation and Remote Control* **65**, 3, pp. 472–482.

Kallenberg, R. H. and Cvjetnicanin, G. D. (1994). *Film into Video: A Guide to Merging the Technologies* (Focal Press).

Kantorovich, L. V. (1937). On Functional Equations, *Uch. zep. LGU* **3**, 7.

Kantorovich, L. V. and Gor'kov, L. P. (1959). Some Functional Equations Arising in the Analysis of Single-Commodity Model of the Economy, *Dokl. Akad. Nauk USSR* **129**, 4, pp. 732–736.

Kantorovich, L. V., Vulikh, B. Z., and Pinsker, A. G. (1950). *Functional Analysis in Semi-ordered Spaces* (Nauka. Moscow-Leningrad).

Kanwal, R. P. (2013). *Generalized Functions: Theory and Applications* (Birkhauser Publ., 3rd edition).

Karaulova, I. V. and Markova, E. V. (2003). On One Optimal Control Problem in the Glushkov Integral Models, in *CD Proceedings of Forth International Conference Inverse Problems: Identification, Design and Control, July 2-6, 2003* (MAI), pp. 45–55.

Karaulova, I. V. and Markova, E. V. (2008). Optimal Control Problem of Development of an Electric Power System, *Differential Equations* **69**, 4, pp. 637–644.

Kárný, M., Böhm, J., Guy, T. V., Jirsa, L., Nagy, I., Nedoma, P., and Tesař, L. (2005). *Optimized Bayesian Dynamic Advising: Theory and Algorithms*

(Springer Publ.).

Kekatos, V. and Giannakis, G. B. (2011). Sparse Volterra and Polynomial Regression Models: Recoverability and Estimation, *IEEE Trans. on Signal Processing* **59**, 12, pp. 5907–5920.

Keller, H. B. (1977). *Numerical Solution of Bifurcation and Nonlinear Problems* (New York: Academpress).

Khawar, J., Zhigang, W., and Chao, Y. (2012). Volterra Kernel Identification of MIMO Aeroelastic System through Multiresolution and Multiwavelets, *Computational Mechanics* **49**, 4, pp. 431–458.

Khromov, A. P. (2006). Integral Operators with Discontinuous Kernel on Piecewise Linear Curves, *Sbornik: Mathematics* **197**, 11, pp. 115–142.

Kiselev, A. V. and Faddeev, M. M. (2000). The similarity problem for non-self-adjoint operators with absolutely continuous spectrum, *Functional Analysis and Its Applications* **34 (2)**, pp. 140–142.

Koh, T. and Powers, E. J. (1985). Second-order Volterra Filtering and its Application to Nonlinear System Identification, *IEEE Acoust. Speech Sig. Proc.* **6**, pp. 1445–1455.

Kokaram, A., Bornard, R., Rares, A., Sidorov, D., Chenot, J.-H., Laborelli, L., and Biemond, J. (2002). Robust and Automatic Digital Restoration System: Coping with Reality, in *International Broadasting Convention 2002, Proc. IBC* (Amsterdam, The Netherlands), pp. 405—411.

Kokaram, A. C. (1998). *Motion Picture Restoration* (Springer-Verlag).

Korba, P. (2007). Real-time Monitoring of Electromechanical Oscillations in Power Systems: First Findings, *Generation, Transmission and Distribution, IET* **1**, 1, pp. 80–88.

Korba, P., Larsson, M., and Rehtanz, C. (2003). Detection of Oscillations in Power Systems Using Kalman Filtering Techniques, in *Proceedings of IEEE Conference on Control Applications, 23-25 June 2003* (IEEE, Washington, D.C.), pp. 183–188.

Korenberg, M. (1988). Identifying Nonlinear Difference Equation and Functional Expansion Representations: the Fast Orthogonal Algorithm, *Ann. Biomed. Eng.* **16**, pp. 123–142.

Kreyszig, E. (1978). *Introductory Functional Analysis* (The Wiley and Sons Publ.).

Kulhavý, R. and Kraus, F. J. (1996). On Duality of Regularized Exponential and Linear Forgetting, *Transactions of the American Mathematical Society* **32**, pp. 1403–1415.

Kundur, P. (1994). *Power System Stability and Control* (McGraw Hill, NY).

Kurbatsky, V., Sidorov, D., Spiryaev, V., and Tomin, N. (2011a). Neural Networks Approach to Nonstationary Time Series Forecast based on the Hilbert-Huang Transform, *Automation and Remote Control* **72**, 7, pp. 1405—1414.

Kurbatsky, V., Sidorov, D., Tomin, N., and Spiryaev, V. (2011b). Hybrid Model for Short-Term Forecasting in Electric Power System, *J. of Machine Learning and Computation* **2**, 5, pp. 138–147.

Kurbatsky, V., Tomin, N., Sidorov, D., and Spiryaev, V. (2012). The Novel Approach for Power Flow Prediction based on the Marginal Hilbert Specrum and ANN, in *Proceedings of 5th Intl Conference on Liberalization and Modernization of Power Systems, Irkutsk, Russia, August 6-10, 2012 (CD)* (ISEM SB RAS, Irkutsk, Russia), pp. 1–8.

Kurbatsky, V., Tomin, N., Spiryaev, V., Leahy, P., and Zjukov, A. (2014). Power System Parameters Forecasting Using Hilbert-Huang Transform and Machine Learning, *arXiv* **1404.2353**, pp. 1–15.

Kurbatsky, V. G., Sidorov, D. N., Spyryaev, V. A., and Tomin, N. V. (2011c). Application of Two Stages Adaptive Neural Network Approach for Short-term Forecast of Electric Power System, in *Proc. of the International conference EEEIC'11* (Rome, Italy), pp. 1–6.

Kurbatsky, V. G., Sidorov, D. N., Spyryaev, V. A., and Tomin, N. V. (2011d). The Hybrid Model based on the Hilbert-Huang Transform and Neural Networks for Forecasting Short-Term Operation Conditions of Power System, in *Proc. of the International Conference PowerTech'2011* (Trondheim, Norway), pp. 1–7.

Kurbatsky, V. G., Sidorov, D. N., Spyryaev, V. A., and Tomin, N. V. (2011e). Short-term Forecasting Parameters of EPS for Systems of Operating and Emergency Control, in *"Actual Trends in Development of Power Systems Protection and Automation", 30 May – 3 June 2011. SPb, Russia* (SPb, Russia), pp. 44–45.

Larsen, T. (1993). Determination of Volterra Transfer Function of Non-linear Multi-Port Networks, *Applied Mathematics and Computation* **2**, 21, pp. 107–131.

Ledwich, G. and Palmer, E. (2001). Modal Estimates from Normal Operation of Power Systems, in *Power Engineering Society Winter Meeting, 2001. IEEE*, Vol. 2 (IEEE, Washington D.C.), pp. 1527 – 1531.

Lee, D. (2011). Short-term Prediction of Wind Farm Output Using the Recurrent Quadratic Volterra Model, in *Power and Energy Society General Meeting, 24-29 July 2011* (IEEE PES, Washington, D.C..), pp. 1–8.

Lee, Y. W. and Schetzen, M. (1965). Measurement of the Wiener Kernels of a Non-linear System by Cross-Correlation, *Int. J. Control* **2**, pp. 237–254.

Lichtenstein, L. (1931). *Vorlesungen Uber Einige Klassen Nichtlinearer Integralgleichungen und Integro-differentialgleichungen Nebst Anwendungen* (Berlin, Julius Springer).

Ling, W. M. and Rivera, D. E. (2001). A Methodology for Control-Relevant Nonlinear System Identification using Restricted Complexity Models, *J. Proc. Cont.* **11**, pp. 209–222.

Linz, P. (1985). *Analytical and Numerical Methods for Volterra Equations* (SIAM).

Liu, G. P., Kadirkamanathan, K., and Billings, S. A. (1998). On-line Identification of Nonlinear Systems using Volterra Polynomial Basis Function Neural Networks, *Neural Networks* **11**, pp. 1645–1657.

Lorenzi, A. (2001). *An Introduction to Identification Problems Via Functional Analysis* (Utrecht, VSP).

Lorenzi, A. (2013). Operator Equations of the 1st Kind and Integro-Differential

Equations of Degenerate Type in Banach Spaces and Applications to Integro-Diffrential PDE's, *Eurasian Journal of Mathematical and Computer Applications* **1**, 2, pp. 50–75.

Louppe, G., Wehenkel, L., Sutera, A., and Geurts, P. (2013). Understanding Variable Importances in Forests of Randomized Trees, *Advances in Neural Information Processing Systems* **1**, 26, pp. 431—439.

Lyapunov, A. M. (1906). *Notes of Academy of Sciences* (Academiya Nauk. St.Petersburg).

Magnitsky, N. A. (1983). Asymptotics of the Solution of Volterra Integral Equations of the First Kind, *DAN USSR* **169**, 1, pp. 29–32.

Makarov, V. and Khlobystov, V. (1997). Polynomial Interpolation of Operators, *Journal of Mathematical Sciences* **84**, pp. 1244–1290, doi:10.1007/BF02399122.

Markova, E. V., Sidler, I. V., and Trufanov, V. V. (2011). On Models of Developing Systems and their Applications, *Automation and Remote Control* **72**, 7, pp. 1371–1379.

Markova, E. V. and Sidorov, D. N. (2014). On one Integral Volterra Model of Developing Dynamical Systems, *Automation and Remote Control* **75**, 3, pp. 413–421.

Marmarelis, P. Z. and Marmarelis, V. Z. (eds.) (1978). *Analysis of Physiological Systems: the White-Noise Approach* (Plenum Press Ltd., New York).

Mathews, V. and Sicuranza, G. (2000). *Polynomial Signal Processsing* (New York: Wiley).

Mercer, J. (1909). Functions of Positive and Negative Type, and their Connection with the Theory of Integral Equations, *Philosophical Transactions of the Royal Society of London. Series A, Containing Papers of a Mathematical or Physical Character* **209**, pp. 415–446.

Messina, A. R., Olguin, S. D., Rivera, S. C. A., and Ruiz-Vega, D. (2001). Analytical Investigation of Large-Scale use of Static VAR Compensation to Aid Damping of Inter-Area Oscillations, in *Proceedings of the Seventh International Conference on AC-DC Power Transmission (Conf. Publ. No. 485), 28-30 November 2001* (IEEE, Washington, D.C..), pp. 187–192.

Micke, A. (1989). The Treatment of Integral Equations with Discontinuous Kernels Using Product Type Quadrature Formulas, *Computing* **42**, pp. 207–223.

Miller, B. M. and Rabinovich, E. Y. (2003). *Impulsive Control in Continuous and Discrete-continuous Systems* (NY: Kluwer Academic Publ.).

Moore, G. (1980). The Numerical Threatment of Non-trivial Bifurcation Points, *Numer. Funct. Anal. Optim.* **2**, 6, pp. 441–472.

Németh, J. G., Kollár, I., and Schoukens, J. (2002). Identification of Volterra Kernels using Interpolation, *IEEE Trans. on Instrumentation and Measurement* **4**, 51, pp. 770–775.

Nishijima, Y. and Oster, G. (1964). Moiré Patterns: their Application to Refractive Index and Refractive Index Gradient Measurements, *Jour. of the Optical Society of America* **54**, pp. 1—5.

Nowak, R. D. and Veen, B. D. V. (1994). Efficent Methods for Identification of

Volterra Filter Models, *Signal Processing* **38**, pp. 417–428.

O'Regan, D. (2011). Nonlinear Hammerstein Integral Equations via Local Linking and Mountain Pass, *Rend. Circ. Mat. Palermo* **60**, pp. 357–364.

Oster, G. (1964). *The Science of Moiré Patterns* (Edmund Scientific Co., USA).

Oster, G., Wasserman, M., and Zwerling, C. (1964). Theoretical Interpretation of Moiré Patterns, *Jour. Of the Optical Society of America* **54**, pp. 169—175.

Parker, R. P. and Tummala, M. (1992). Identification of Volterra Systems with a Polynomial Neural Network, in *Proc. 1992 IEEE Int. Conf. Acoust., Speech, Signal Processing*, Vol. 4 (San Francisco, CA, USA), pp. 561–564.

Parker, R. S., Heemstra, D., Doyle, F. J., Pearson, R. K., and Ogunnaike, B. A. (2001). The Identification of Nonlinear Models for Process Control using Tailored Plant-Friendly Input Sequences, *J. Proc. Control* **11**, pp. 237–250.

Pearson, R. K., Ogunnaike, B. A., and Doyle, F. J. (1996). Identification of Structurally Constrained Second-order Volterra Models, *IEEE Trans. on Signal Processing* **44**, 11, pp. 2837–2846.

Pei, S.-C. and Tseng, C.-C. (1994). Two Dimensional IIR Digital Notch Filter Design, *IEEE Trans. on Circuits and Systems II: Analog and Digital Signal Processing* **41**, 3, pp. 227—231.

Petrov, Y. P. and Sizikov, V. S. (eds.) (2005). *Well-posed, Ill-posed, and Intermediate Problems and Applications* (V.S.P. Intl Science).

Plato, R. and Vainikko, G. (1990). On the regularization of projection methods for solving ill-posed problems, *Numer. Math.* **57**, pp. 63–79.

Prochelt, L. (1994). *Proben1 - A Set of Neural Network Benchmark Problems and Benchmark Rules*, Vol. 21 (University of Karlsruhe, Germany).

Pruss, J. (2012). *Evolutionary Integral Equations and Applications* (Birkhauser).

Qingsheng, E. and Zafiriou, E. (1995). Nonlinear System Identification for Control using Volterra-Laguerre Expansion, in *Proc. of American Control Conference, 1995*, Vol. 3 (Seattle, USA), pp. 2195–2199.

Reinhardt, H.-J. (1985). *Analysis of Approximation Methods for Differential and Integral Equations* (Springer, NY).

Rubiolo, M., Stegmayer, G., and Milone, D. (2012). Compressing Arrays of Classifiers using Volterra-Neural Network: Application to Face Recognition, *Neural Computing and Applications* , pp. 1–15.

Rugh, W. J. (1981). *Nonlinear System Theory* (The John Hopkins University Press).

Scherbinin, M. S. (2010). R and D Bulletin of Rosneft JSC, *Power Consumption Optimization for the Turbine Compressor M1 EP-300* **3**, pp. 24–27.

Schetzen, M. (1980). *The Volterra and Wiener Theories of Non-Linear Systems* (New York: Wiley-Interscience).

Schwarz, H. G. (2005). Modernisation of Existing and New Construction of Power Plants in Germany: Results of an Optimisation Model, *Energy Economics* **27**, pp. 113–137.

Shcherbakov, M. A. (1996). Parallel Implementation of Digital Volterra Filters in the Frequency Domain, *Automatic Control and Computer Sciences* **30**, 6, pp. 25–31.

Shilov, G. E. and Gurevich, B. (1978). *Integral, Measure, and Derivative: A Unified Approach* (Dover Publications).

Shwartz, L. (1961). *Méthodes Mathématiques pour les Sciences Physiques* (Paris: Hermann).

Sidorov, D. (1999). *Modeling of Nonlinear Dynamical Systems with the Volterra Series: Identification and Applications* (ISEM SB RAS).

Sidorov, D. (2002). *Modelling of Non-linear Dynamic Systems by Volterra Series. In "Attractors, Signals, and Synergetics"* (Pabst Science Publ. USA-Germany).

Sidorov, D. (2011a). On Impulsive Control of Nonlinear Dynamical Systems Based on the Volterra Series, in *10th IEEE International Conference on Environment and Electrical Engineering, EEEIC* (Rome, Italy), pp. 1–3.

Sidorov, D. (2011b). Volterra Equations of the First Kind with Discontinuous Kernels in the Theory of Evolving Systems Control, *Studia Informatica Universalis. Paris: Hermann Publ.* **9**, 3, pp. 135–146.

Sidorov, D. and Kokaram, A. (2002). Suppression of Moiré Patterns via Spectral Analysis, in *Visual Communications and Image Processing 2002, Proc. SPIE 4671* (San Jose, USA), pp. 895—906.

Sidorov, D., Panasetsky, D., and Šmídl, V. (2010a). Non-stationary Autoregressive Model for An-line Detection of Inter-area Oscillations in Power Systems, in *Innovative Smart Grid Technologies Conference Europe (ISGT Europe), 2010 IEEE PES, 11-13 Oct. 2010* (IEEE PES, Washington, D.C..), pp. 1–5.

Sidorov, D., Tynda, A., and Muftahov, I. (2014). Numerical Solution of Weakly Regular Volterra Integral Equations of the First Kind, *arXiv* **1403.3764v2**, pp. 1–8.

Sidorov, D., Wei, W. S., Vasilyev, I., and Salerno, S. (2008). Automatic defects classification with p-median clustering technique, in *Control, Automation, Robotics and Vision, 2008. ICARCV 2008. 10th International Conference on*, pp. 775 –780, doi:10.1109/ICARCV.2008.4795615.

Sidorov, D. N. (2013). Solution to Systems of Volterra Integral Equations of the First Kind with Piecewise Continuous Kernels, *Russian Mathematics (Transl. from Izvestia VUZov)* **57**, 1, pp. 62–72.

Sidorov, D. N., Grishin, Y. A., and Šmídl, V. (2010b). On-line Detection of Inter-area Oscillations using Forgetting Approach for Power Systems Monitoring, in *IEEE Proc. of The 2nd International Conference on Computer and Automation Engineering (ICCAE), 26–28 Feb. 2010, Singapore*, Vol. 3 (IEEE, Washington, D.C.), pp. 292–295.

Sidorov, D. N. and Sidorov, N. A. (2011a). Generalized Solutions in Problem of Modeling of Nonlinear Dynamic Systems with the Volterra Polynomials, *Automations and Remote Control* **6**, pp. 127–132.

Sidorov, D. N. and Sidorov, N. A. (2012). Convex Majorants Method in the Theory of Nonlinear Volterra Equations, *Banach J. Math. Anal.* **6**, 1, pp. 1–10.

Sidorov, N., Sidorov, D., and Leontiev, R. (2012). Successive Approximations to the Solutions to Nonlinear Equations with a Vector Parameter in a

Nonregular Case, *Journal of Applied and Industrial Mathematics* **6**, pp. 387–392, doi:10.1134/S1990478912030143.

Sidorov, N. A. (1997). An n-step Iteration Method in Branching Theory for Solutions to Nonlinear Equations, *Siberian Mathematical Journal* **38**, 2, pp. 330–341.

Sidorov, N. A., Falaleev, M. V., and Sidorov, D. N. (2006). Generalized Solutions of Volterra Integral Equations of the First Kind, *Bull. Malays. Math. Soc.* **29**, 2, pp. 1–5.

Sidorov, N. A., Loginov, B. V., Sinitsyn, A. V., and Falaleev, M. V. (2002). *Lyapunov-Schmidt Methods in Nonlinear Analysis and Applications* (Kluwer Academic Publ.).

Sidorov, N. A. and Sidorov, D. N. (2006). Generalized Solutions to Integral Equations in the Problem of Identification of Nonlinear Dynamic Models, *Differential Equations* **42**, 9, pp. 1312–1316.

Sidorov, N. A. and Sidorov, D. N. (2011b). Small Solutions of Nonlinear Differential Equations near Branching Points, *Russian Mathematics (Transl. from Izvestia VUZov)* **55**, 5, pp. 43–50.

Sidorov, N. A., Sidorov, D. N., and Krasnik, A. V. (2010c). Solution of Volterra Operator-integral Equations in the Nonregular Case by the Successive Approximation Method, *Differential Equations* **46**, 6, pp. 882–891.

Sidorov, N. A., Sidorov, D. N., and Trufanov, A. V. (2007). Existence and Structure of Solution of Integral-Functional Volterra Equations of the First Kind, *Bulletin of Irkutsk State University: Mathematics* **1**, 1, pp. 267–274.

Sidorov, N. A. and Trenogin, V. A. (1978). Regularization of Simple Solutions to Nonlinear Equations in the Neighborhood of Branch Point, *Siberian Mathematical Journal* **20**, 1, pp. 180–183.

Sidorov, N. A. and Trufanov, A. V. (2009). Nonlinear Operator Equations with a Functional Perturbation of the Argument of Neutral Type, *Differential Equations* **45**, 12, pp. 1840–1844.

Small, M. (2005). *Applied Nonlinear Time Series Analysis, A*, Vol. 52 (World Scientific Series on Nonlinear Science).

Smith, H. (2010). *An Introduction to Delay Differential Equations with Applications to the Life Sciences* (New York: Sringer).

Smola, A. J. and Schölkopf, B. (2004). A Tutorial on Support Vector Regression, *Statistics and Computing* **14**, pp. 199–222.

Solow, R. M. (1969). Investment and Technical Progress, *Mathematical Methods in the Social Sciences* , pp. 89–104.

Soni, A. S. (2006). *Control-Relevant System Identification Using Nonlinear Volterra and Volterra-Laguerre Models* (PhD Thesis: Pittsburg Univ., Pittsburg, USA).

Soni, A. S. and Parker, R. S. (2004). Control Relevant Identification for Third-Order Volterra systems: a Polymerization Case Study, in *Proc. of American Control Conference, 2004*, Vol. 5 (Boston, USA), pp. 4249–4254.

Spiecker, S. and Weber, C. (2011). Integration of fluctuating renewable energy 2014: A german case study, in *Power and Energy Society General Meeting, 2011 IEEE*, pp. 1–10.

Spiryaev, V. A. (2006). Application of the Product Integration Method to Identification of the Volterra Kernels, *Sistemnye issledovania v energetike. ISEM SO RAN.* **36**, pp. 215–221.

Sveshnikov, A. G., Alshin, A. B., Korpusov, M. O., and Pletnev, Y. D. (2007). *Linear and Nonlinear Sobolev Equations* (Fizmatlit, Moscow).

Sviridyuk, G. A. and Fedorov, V. E. (2003). *Linear Sobolev Type Equations and Degenerate Semigroups of Operators* (Utrecht: VSP).

Tairov, E. A. (1989). Nonlinear Modeling of the Dynamics of Heat Transfer in a Channel with Single Phase Coolant, *Izv. AN SSSR: Energetika i transport* **1**, pp. 150–156.

Tairov, E. A. and Zapov, V. V. (1991). Integral Model of the Nonlinear Dynamics of Steam-generating Channel based on Analyical Solutions, *Issues of Nuclear Sc. and Tech. Physics of Nuclear Reactors* **3**, pp. 14–20.

Takasaki, H. (1970). Moiré Topography, *Applied Optics* **9**, 6, pp. 1467—1472.

Theocaris, P. S. (1963). *Moiré Fringes in Strain Analysis* (Pergamon Press, UK).

Tikhonov, A. N. and Arsenin, V. Y. (1977). *Solutions of Ill-posed Problems* (Wiley, New York (Trans. from Russian)).

Torokhti, A. and Howlett, P. (2007). Computational methods for modeling of nonlinear systems, , p. 322.

Toshio, U., Imsik, H., and Terashi, T. (1977). *Some notes on linear Fredholm integral equations of the first kind* (Research Inst. of Science and Technology, Nihon Univ., Tokyo).

Trénoguine, V. A. (1985). *Analyse fonctionnelle* (Éd. Mir).

Tsypkin, Y. Z. and Popkov, Y. S. (1973). *Theory of Nonlinear Pulse Systems* (Nauka).

Vaidyanathan, P. P. (2007). *The Theory of Linear Prediction*, Synthesis Lectures on Signal Processing (Morgan and Claypool Publ.).

Vainberg, M. M. and Trenogin, V. A. (1974). *Theory of Branching of Solutions of Non-linear Equations* (Noordhoff).

Vapnik, V. (1998). *Statistical Learning Theory* (John Wiley and Sons, Inc., New York).

Vladimirov, V. S. (1979). *Generalized Functions in Mathematical Physics* (Mir).

Volmir, A. S. (1967). *Stability of Deformable Systems* (Nauka, Moscow).

Volterra, V. (2005). *Theory of Functionals and of Integral and Integro-Differential Equations* (New York: Dover Publ.).

Voropai, N. I., Gamm, A. Z., Glazunova, A. M., Etingov, P. V., Kolosok, I. N., Korkina, E. S., Kurbatsky, V. G., Sidorov, D. N., Spiryaev, V. A., Tomin, N. V., Zaika, R. A., and Bat-Undraal, B. (2013). *Application of Meta-Heuristic Optimization Algorithms in Electric Power Systems* (IGI Global).

Voropai, N. I., Kolosok, I. N., Kurbatsky, V. G., Etingov, P. V., Tomin, N. V., Korkina, E. S., and Paltsev, A. S. (2010). Intelligent Coordinated Operation and Emergency Control in Electric Power Systems, in P. M. Ferrera (ed.), *Proc. IFAC Conference on Control Methodologies and Technology for Energy Efficiency (CMTEE-2010)* (Vilamoura, Portugal), pp. 189–203.

Voskoboinikov, Yu. E. (2007). A combined nonlinear contrast image reconstruction algorithm under inexact point-spread function, *Optoelectronics,*

Instrumentation and Data Processing **43**, 6, pp. 489–499, doi:10.3103/ S8756699007060015.

Šmídl, V. and Quinn, A. (2006). *The Variational Bayes Method in Signal Processing* (Springer. Series: Signals and Communication Technology).

Wiener, D. D. and Spina, J. (1980). *Sinusoidal Analysis and Modeling of Weakly Nonlinear Circuits* (Van Nostrand, New York).

Wiener, N. (ed.) (1958). *Nonlinear Problems in Random Theory* (Wiley Ltd., New York).

Wilde, D. J. (1964). *Optimum Seeking Methods* (Englewood Cliffs, Prentice- Hall Inc.).

Wu, J. (2006). Estimation Electro-Mechanical Parameters using Frequency Measurements, in *Materials of CIGRE Regional Conference "Monitoring of Power System Dynamics Performance"* (CIGRE, 21, Moscow), pp. 12–17.

Yaroslavsky, L. P. and Eden, M. (1996). *Fundamentals of Digital Optics* (Birkhauser, Boston).

Zavalishin, S. T. and Sesekin, A. N. (1997). *Dynamic Impulse Systems: Theory and Applications* (Dordrecht: Kluwer Academic Publ.).

Zeidler, E. (1986). *Nonlinear Functional Analysis and its Applications I* (Springer Verlag).

Zhang, Q., Suki, B., Westwick, D. T., and Lutchen, K. R. (1988). Factors Affecting Volterra Kernel Estimation: Emphasis on Lung Tissue Viscoelasticity, *Ann. Biomed. Eng.* **26**, pp. 103–116.

Zhijian, L., Hongchun, S., and Jilai, Y. (2008). Analytical Investigation of Large-Scale use of Static VAR Compensation to Aid Damping of Inter-Area Oscillations, in *International Symposium on Intelligent Information Technology Application Workshops, Los Alamitos, CA, USA* (IEEE Computer Society, Washington, D.C..), pp. 205–208.

Acronyms

ANN Artificial Neural Networks
AR Auto-regression
ARX Auto-regressive Model with Exogenous Inputs
CSRS Constant-switching-pace Symmetric Random Sequences
CPDF Cumulative Probability Distribution Function
EMD Empirical Mode Decomposition
EPS Electric Power Systems
FACTS Flexible Alternating Current Transmission System
FFT Fast Fourier Transform
GA Genetic Algorithm
GRNN General Regression Neural Networks
GWN Gaussian White Noise
HHT Hilbert Huang Transform
HT Hilbert Transform
IMF Intrinsic Mode Functions
KF Kalman Filter
LMS Least Mean Squares
MAE Mean Absolute Error
MAPE Mean Absolute Percentage Error
MIMO Multiple Input Multiple Output
MISO Multiple Input Single Output
MLP Multi-Layer Perceptron
MPC Model Predictive Control
NGIS Neural Genetic Input Selection
NHIE Nonlinear Hammerstain Integral Equation
PDF Probability Distribution Function
PEV Prediction Error Variance

PSS Power System Stabilizer
RBF Radial Basis Function
RBS Random Binary Sequence
RKHS Reduced Kernel Hilbert Space
RLS Recursive Least Squares
RMSE Root Mean Squared Error
PMU Phasor Measurement Units
SA Simulated Annealing
SISO Single Input Single Output
SLAE System of Linear Algebraic Equations
SNR Signal-to-Noise Ratio
SVC Static Var Compensator
SVM Support Vector Machine
SVR Support Vector Regression
TCSC Thyristor Controlled Series Compensator
VIE Volterra Integral Equation
VCM Vintage Capital Model
Volterra-NN Volterra Neural Network

Index